D0371201

Power Surge

OTHER NORTON/WORLDWATCH BOOKS

Lester R. Brown et al.
State of the World 1984
State of the World 1985
State of the World 1986
State of the World 1987
State of the World 1988
State of the World 1989
State of the World 1990
State of the World 1991
State of the World 1992
State of the World 1993
State of the World 1994
Vital Signs 1992
Vital Signs 1993
Vital Signs 1994

ENVIRONMENTAL ALERT SERIES

Lester R. Brown et al.
Saving the Planet

Alan Thein Durning
How Much Is Enough?

Sandra Postel
Last Oasis

Lester R. Brown
Hal Kane
Full House

POWER SURGE

Guide to the Coming Energy Revolution

Christopher Flavin
Nicholas Lenssen

The Worldwatch Environmental Alert Series
Linda Starke, Series Editor

W · W · NORTON & COMPANY
NEW YORK   LONDON

Worldwatch Database Diskette

The data from all graphs and tables contained in this book, as well as from those in all other Worldwatch publications of the past year, are available on diskette for use with IBM-compatible or Macintosh computers. This includes data from the Vital Signs *series of books,* Worldwatch Papers, World Watch *magazine, and the Environmental Alert series of books. The data are formatted for use with spreadsheet software compatible with Lotus 1-2-3 version 2, including all Lotus spreadsheets, Quattro Pro, Excel, SuperCalc, and many others. For IBM-compatibles, both 3½- and 5¼-inch diskettes are supplied. To order, send check or money order for $89.00, or credit card number and expiration date (Visa and MasterCard only), to Worldwatch Institute, 1776 Massachusetts Ave. NW, Washington, DC 20036 (tel: 202-452-1999; fax: 202-296-7365).*

Contents

Acknowledgments

After following the ups and downs in the energy field since the late seventies, we embarked on this book with some trepidation about how much new there was to say. After all, many people seemed convinced that the exciting years in energy policy were behind us and that the forces of the status quo had retaken control of the global energy system.

Two years later, we write the final lines of *Power Surge* in a far different frame of mind—excited about the embryonic but far-reaching changes in energy technologies and policies we have discovered in disparate corners of the globe, and genuinely optimistic about the potential for a fundamental reshaping of the global energy system during the next few decades.

From Sri Lankan villagers installing solar electric systems on their rooftops, to Swiss tinkerers building light-weight electric cars, to large corporations laying wind farms across the Great Plains of North America, an energy revolution is under way, leading to a system that is more efficient, decentralized, and participatory—and far less taxing on the global environment. Indeed, we are firmly convinced of the practicality of achieving an energy economy that does not pollute our cities, sicken our lakes and forests, or cause long-term changes in the earth's climate.

Most of the technologies and policies needed to achieve these important goals are identified in *Power Surge*. The speed with which the shifts occur hinges largely on politics—how quickly the heavy subsidies and monopoly markets of the past are done away with, and how rapidly a more competitive and open energy system emerges. In the last five years, decades-old energy policies have been reversed in many countries, and in today's world, successful innovations can spread rapidly from one country to another. So, even in energy politics, we are increasingly optimistic.

During the past decade we have developed a network of extraordinarily knowledgeable colleagues around the world on whom we rely for the latest developments and most innovative insights. This loose group of experts and friends has engaged us on many occasions in lengthy conversations about the issues they work on, and responded promptly to urgent requests for information. Among those we relied on, including many who reviewed portions of this manuscript are: Dennis Anderson, Robert Annan, Hans Bjerregård, Gerald Braun, Bill Browning, Jim Caldwell, Ralph Cavanagh, Armond Cohen, Øystein Dahle, Michael Davis, Sally de Becker, Dennis Elliott, S. David Freeman, Uwe Fritsche, Howard Geller, Michael German, Paul Gipe, David Hall, Jan Hamrin, Robert A. Hefner III, John Hemphill, Robert Hill, Richard F. Hirsh, Eric Hirst, Harold Hubbard, Fred Julander, René Karottki, Will Keepin, Keith Kozloff, David Lapp, Eric Larson, Daniel Lashof, Ken Lay, Reinhard Loske, Amory Lovins, Ragnar Lundqvist, Robert Lynette, Paul MacCready, Paul Maycock, William McDonough, Alden Meyer, David Mills, David Moskovitz, Urs Muntwyler, Joan Ogden, Donald Osborn, Richard Ottinger, David Penn, David Pimentel, Kevin Porter, Harvey Sachs, Lee Schipper, Bruce Stram, Steven Strong, Tom Surek, Randy Swisher, Steven Ternoey, John Thornton, Hardin Tibbs, Eric Toler, Michael Walsh, Carl Weinberg, Stephen Wiel, Neville Williams, Robert H. Williams, and Koichi Yamanashi.

We also benefited from in-house reviews of the manuscript by our Worldwatch colleagues, including Lester Brown, Hilary French, Hal Kane, Jim Perry, Sandra Postel, and Michael Renner. And we want to express special thanks to the boards and staff of the two major funders of this book—The Energy Foundation in San Francisco and the Joyce Mertz-Gilmore Foundation in New York. Their dedication to the cause of achieving a more sustainable energy system has been invaluable to numerous nongovernmental organizations.

In researching and writing this book at the Worldwatch Institute, we have relied on an exceptional support system that includes efficient and professional communications, publications, and administrative departments that keep the gears turning and ensure that our work reaches a broad audience. We have also depended on the close support of two capable assistants— Suzanne Clift and Kari Whealen. On the research side,

David Roodman was a full partner almost from the start. His extraordinary analytical abilities, particularly his capacity for assessing complex quantitative issues, and his penchant for detecting flaws in our logic were invaluable. And we enjoyed the able assistance of intern Michael Scholand, an engineering student at Tufts University. Linda Starke, the editor of the Environmental Alert Series, has as usual done a fine job of editing our sometimes clumsy prose and keeping us almost on schedule.

Finally, on a personal note, we want to thank Hilary French and Maureen Lenssen for their patience and support during a long and often arduous process.

Christopher Flavin
Nicholas Lenssen

Worldwatch Institute
1776 Massachusetts Ave., N.W.
Washington, D.C. 20036

July 1994

Foreword

This is the fifth book in the Worldwatch Environmental Alert Series. Like the third one, which was on water, it deals with something that people in the industrial world now take for granted—in this case, energy. We turn a switch, and expect a light to go on or a microwave oven to cook our dinner. Few of us think about the source of the electricity we count on, or the cost to the earth of generating it.

But many Americans do remember when energy was the number one topic of conversation. In the seventies, long gas lines made us reconsider our assumptions about endless supplies, at least of petroleum. Spurred by two oil crises during that decade, enormous strides have been made in efficiency—in doing more with less energy. Yet it is easy for complacency to slip back in and for rising consumption levels to wipe out the gains achieved through improving energy productivity levels. In 1993, the United States imported 6.7 million barrels of oil, a record 49 percent of what Americans used that year.

In *Power Surge*, Christopher Flavin and Nicholas Lenssen look at the need for even more improvements in energy productivity as part of the revolution that must take place if we are to meet the world's needs for heat, light, transportation, and other basic services. Likewise, the book describes the host of new energy-producing technologies that will allow renewable sources to power the global economy. They conclude that "the coming enegy revolution will have profound effects on the way all of us work and live, and on the health of the global environment on which we depend."

They come to encouragingly different conclusions about the scope for change than many other analysts who have considered the world's energy future. That is because corporations and governments that paint a picture of global energy use in the next 30–50 years usually do so by looking through a rearview mirror. As Chris and Nick point out, the possibility of major technological advances is rarely considered by these anaylsts, and change is seen as an incremental process along well-worn paths.

Technological and institutional changes of the scale discussed in *Power Surge* are seldom imagined by other energy analysts, but there is ample proof that they can occur. The authors draw telling parallels with the dramatic changes in computers and telecommunications. In one case, technology has revolutionized the industry; in the other, government-induced change has transformed the way services are delivered.

Similar revolutions can and must lie ahead in the energy used in industrial and developing countries and in the way we receive the services that energy provides. The environmental health of the planet and the economic health of our societies depend on it.

Power Surge provides details of one key part of the picture of a sustainable society outlined in the first book in the Environmental Alert Series: *Saving the Planet.* The second volume was *How Much Is Enough?*, which looked at consumption patterns and the changes that are needed in rich and poor societies alike. The third, *Last Oasis,* considered water scarcity and the similar revolution in efficiency needed in that sector. And in *Full House*, Lester Brown and Hal Kane reassessed the earth's population carrying capacity in light of the pressures on food production systems and on our ability to meet other needs of rapidly growing populations.

We hope that you find these studies useful additions to a growing library on sustainability, which provides both challenges and opportunities for us all in the years ahead.

Linda Starke, Series Editor

I

Pressures for Change

1

Power Shift

The best way to predict the future is to invent it.[1]

Alan Kay
Personal Computer Pioneer

There is little doubt that we live in an age of rapid and accelerating change. Indeed, in today's world everything at times seems to be in flux. We live vastly different lives than our parents did, and popular literature is filled with speculation about how the next decade will be markedly different from the current one.

In almost every corner of the globe, new technologies are transforming the way we live—from jet travel to frost-free refrigeration—in ways that are sometimes positive, sometimes negative, and often ambiguous. Little more than a decade after personal computers were introduced, they have become ubiquitous, changing the way that tens of millions of people work. During the same period, television has transported hundreds of millions of people in developing countries from self-sufficient isolation to membership in the global village, viewing world events live on CNN.

Social and political mores are changing as well. From South Africa to southern Mexico, people in dozens of countries are demanding participatory democracy and human rights from once-authoritarian political systems, and dismantling many of the inefficient state enterprises that had dominated their economies. In 1989, the people of Eastern Europe suddenly and decisively removed four-decade-old communist regimes that had failed to deliver democratic political rights, economic prosperity, or a healthy environment. Two years later, the Soviet Union was not just transformed but abolished, something few political "experts" had assumed could ever happen.

The global environment is also changing rapidly—and unpredictably—spurred by unprecedented growth in the number of human beings and an even more rapid expansion in the number of factories, buildings, and automobiles that those people use. The eighties brought a number of disturbing developments: discovery of a rapidly expanding ozone hole over Antarctica, evidence of unprecedented loss of plant and animal species, new links between synthetic chemicals and a host of human diseases, declining yields from scores of important fisheries, and increasingly solid evidence that we and our children face a future of rapid and disruptive climate change. Concern about the local and global environment has rapidly become a potent political force, and the reaction to that concern has begun to influence economic trends.

As the end of the millennium approaches and the pace of change accelerates, our collective curiosity about the future has heightened. Hundreds of articles and books have explored the impact of everything from new information technologies to the rise of genetic engineering and the breakdown of traditional social structures. Scores of theories have been proposed about how each of these could shape human society.

Yet thinking about the world's energy future is surprisingly bereft of the kind of vision found in so many other areas. Outside of a few pockets of innovation, planners and analysts at the world's leading energy institutions seem trapped in the stagnation and confusion that began with repeated failures in the seventies and eighties to develop a viable energy strategy.

Indeed, a review of official energy forecasts and expert studies reveals a rather shocking consensus that during the next few decades only minor changes will occur, yielding a slightly more efficient, marginally cleaner version of today's fossil-fuel-based energy economy. According to the experts, any other kind of energy future would be impossibly expensive and impractical.[2]

What is missed in the insular world of large energy institutions are the forces of change that come from the outside. In earlier decades, energy systems have been profoundly shaped by the same kind of technological, political, and economic forces now affecting the world in so many ways. In recent years, new technologies for harnessing, moving, and using energy, new public expectations for participation in energy decision making, and a series of environmental crises have become potent forces in the energy marketplace. Each is powerful in its own right, and together they have the potential to reshape the global energy economy.

Conventional wisdom is sometimes reliable when anticipating smooth trends, but it almost never anticipates major discontinuities. In fact, energy forecasters have had a track record of nearly unblemished failure during the past two decades. Reports by leading institutions in the early seventies overestimated the amount of nuclear power the world would now be using by a factor of six, while studies in 1980 said oil would cost $100 a barrel by the early nineties.[3]

Although corporations and governments are now using more powerful computers and have adjusted their assumptions to account for earlier mistakes, they still seem to be looking at the future through a rearview mirror. The possibility of major technological advances is rarely considered, and change is usually looked at as an incremental process that follows well-worn paths. To make matters worse, many such assessments are as much a matter of political consensus building as they are serious analysis—telling decision makers what they want to hear, not what they need to know.

The consensus view of the world's energy future, as outlined by the World Energy Council and other organizations, can best be described as status quo-plus. It anticipates an energy economy marked by continuing heavy dependence on fossil fuels, made possible by a gradual shift away from conventional oil toward lower-grade (and more abundant) fossil resources such as lignite and oil shale, in order to avoid the need to make significant technological or policy changes. In this vision, liquid fuels will continue to run the world's automobiles, trucks, and planes while coal-derived electricity provides most of the rest of the massive amounts of energy that will be needed. The experience of the past two decades has forced most analysts to concede that energy will be used more efficiently in the future, but they assume that even these improvements will be negated by massive growth in the energy needs of an expanding world population.[4]

For all their thousands of variables and carefully as-sembled expert groups, what is missing in most studies of the world energy future is the one thing that should be most obvious: the energy trends they describe are not sustainable—either economically or environmentally—and will almost certainly be derailed by an array of forces now intensifying, from the political changes sweeping Russia to the ecological changes threatening the health of millions. Equally misleading, most energy studies neglect the potential for rapid technological change of the sort now driving advances in electronics, telecommunications, biotechnology, and many other fields. Few even question the assumption that it will take decades for new energy technologies to be deployed on a significant scale.

In a world changing as rapidly as ours is today, many plausible scenarios could be constructed, but a 30-year continuation of the status quo is one of the least likely. Indeed, rapid change of many kinds is already taking place, though seeing it requires delving beneath the broad energy statistics that preoccupy most analytical efforts.

One area in which the forces of change have become visible in the past few years is in the host of more energy-efficient technologies that have emerged. From light bulbs to refrigerators, many new energy-using technolo-gies are at least 75 percent more efficient than the cur-rent standard. Even in the power industry, which has sought to improve the efficiency of its equipment for a century, the plants that opened in the early nineties are 50 percent more efficient than a decade earlier. Such advances suggest that government and international agencies are still overstating future levels of energy use.[5]

The energy landscape has also been altered by the recent emergence of natural gas as the most rapidly growing major energy source. Driven by its environmental advantages and global abundance, gas now seems likely to become the dominant global fuel of the early twenty-first century. This would allow considerable displacement of oil and coal in the near future, and lead to an array of much more efficient and decentralized energy conversion and storage systems. The trend to gas would continue the move toward cleaner, more versatile fuels that has marked the last century, and counters the conventional view that the world will rely on increasingly dirty fuels in the coming decades.

In addition, recent developments are turning wind power, solar energy, and a host of other renewable resources into economically viable energy options. Some 20,000 wind turbines are already spread across the mountain passes of California and the northern plains of Europe, while tens of thousands of Third World villagers are getting their electricity from solar cells. Solar and wind energy are far more abundant than any of the fossil energy resources in use today, and declining costs are expected to make them fully competitive in the near future.[6]

The nineties are marked by another unanticipated technological development: viable alternatives to the gasoline-powered internal combustion engine. Lightweight hybrid-electric vehicles made of synthetic materials and run on devices such as gas turbines, fuel cells, and flywheels are about to emerge from engineering labs around the world. With fuel economies that are three to four times the current average and emissions of air pollutants that may be a mere 5 percent of currently permitted levels, these revolutionary new cars, trucks, and buses appear likely to enter the commercial market by the end of this decade, ushering in an era when automobiles can be refueled at home from the local electric or gas system.[7]

The accelerating pace of change in energy systems demands a new paradigm for assessing and managing future developments. What passes for energy analysis today is dominated by a preoccupation with resource supplies and the geopolitics of the Persian Gulf, leaving unquestioned the assumptions that we will stay hooked on oil until it is virtually gone and that coal's role must expand simply because of its abundance.

If analysts had held a similarly blinkered view of other sectors, we would still be driving around in buggies and writing on typewriters. The world never ran out of either hay or paper. Rather, people discovered better means of accomplishing things more conveniently and economically. The energy sector is no exception. The age of oil was ushered in at the turn of the century less by the discovery of petroleum, which had been found much earlier, than by the development of a practical internal combustion engine that made oil useful.

As experts consider the future of the energy economy, they may learn more by studying the electronics revolution of the late twentieth century than by applying the geopolitical and geophysical framework of conventional energy analysis. Most new energy systems will be affected by a range of rapidly evolving technologies, many of them incorporating the latest electronics as a way of raising efficiencies and lowering costs. A variety of stronger, lighter, more versatile synthetic materials will also be applied to everything from wind turbine blades to car frames.

The pace of change will be influenced by the fact that the most important new energy technologies are relatively small devices that can be mass-produced in factories—a stark contrast to the huge oil refineries and power plants that dominate the energy economy of the late twentieth century. The economies of mass manufacturing will quickly bring down the cost of the new technologies, and ongoing innovations will be rapidly incorporated in new products, in much the way that today's consumer electronics industry operates.

Even as the transition to new energy sources begins to unfold in the years ahead, societies may well find that they need to push the pace of change even faster and to accelerate the transition to different systems. Most immediately, the growing dependence of the world on Middle Eastern oil will raise the risk of another oil crisis in the late nineties and beyond. Although oil can be replaced by other fossil fuels, which could carry us at least another 100 years, the more profound challenge that will shape the world energy system is stabilizing the earth's atmosphere. The accumulating burden of greenhouse gases that flow from the fossil fuel economy is now putting the global climate at risk, not only for ourselves but for future generations. Although local air pollution and a range of other environmental problems stemming from fossil fuel use are sometimes amenable to simple technical fixes, climate stabilization is not.

Scientists now estimate that the world will need to cut global carbon emissions to at least 60 percent below prevailing levels in order to stabilize atmospheric carbon dioxide at current concentrations. By contrast, the International Energy Agency now projects a nearly 50-percent increase in emissions between 1990 and 2010—most of it in the Third World, where per capita carbon emissions currently range from one twentieth to one fifth the level in industrial countries. To avoid the risk of potentially catastrophic climate shifts in the middle of the next century, when the human economy is expected to be several times larger, the world needs to achieve a rate of carbon emissions per dollar of gross world product that is roughly one tenth the current level. This essentially means an end to the fossil-fuel-based energy economy as we know it.[8]

The challenge will be compounded by the burgeoning energy needs of developing countries, which already have more than 4 billion people—a figure that is projected to reach more than 6 billion in 2020. They currently have more than three fourths of the world's population, but use just over one third of the world's energy—a per capita level that averages less than one eighth that in industrial countries. Third World energy use has doubled since 1970, and is generally projected to double again in the next 15 years and expand sixfold by 2050.[9]

If trends such as these were to raise per capita energy use in developing countries to even one fourth the current industrial-country level by 2025, this by itself would increase total world energy use by 60 percent. Yet many Third World cities already have air quality problems not seen since the early stages of the Industrial Revolution. If developing countries continue to pursue the energy path blazed by industrial ones in decades past—depending heavily on oil and coal—the combined economic and environmental burdens will begin to undermine their development efforts.[10]

To meet the needs of poor nations and of the global environment, incremental change in today's energy systems is clearly not enough. Instead, the world will need to forge a new energy path, making a gradual transition to an entirely different energy system—one that ultimately relies on renewable sources of energy.

One of the most perplexing questions that remains is how, in a world that does not rely on oil or coal, energy would be stored and transported. Although electricity will undoubtedly play a role, hydrogen is probably the best candidate to be the major energy carrier of the mid-twenty-first century. Pointed to as a potential energy carrier by author Jules Verne more than a century ago, hydrogen is both the lightest of the chemical elements and the simplest possible fuel. It can be manufactured directly from water, using solar or wind energy, or from biomass.[11]

Already, hydrogen is used as an industrial chemical and fuel and as a rocket propellant in space programs. Hydrogen technologies continue to advance rapidly, and the barriers to widespread use of this fuel are falling. Using hydrogen, it will be possible to carry energy via pipeline in the way that natural gas is today—and do so for much less than it now costs to move electricity. Several decades from now, hydrogen may be piped from the windy Great Plains of North America to the eastern seaboard, and from the deserts of western China to the populous coastal plain. At the other end of the pipe, it can be used directly in super-efficient factory equipment and in a variety of household appliances, including fuel cells that allow households to generate their own electricity.

Hydrogen will contribute to the evolution of a more efficient, dispersed, and elegant energy system. In the coming decades, for example, buildings could have solar collector systems on their rooftops and facades, used both for capturing solar heat and producing electricity. The roofs of industrial factories may also have solar gathering systems to power the operations inside. Meanwhile, a new generation of storage devices could allow consumers to store electricity at home so that it is available when needed.

Such technologies may accelerate the restructuring of electric and gas utility systems. Electricity supply appears poised to become a competitive wholesale business while local distribution systems focus on providing energy services—even helping homeowners finance insulation and rooftop solar systems. Finally, transportation is unlikely to remain static. Motor vehicles will almost certainly become more efficient and cleaner, running on natural gas, hydrogen, and electricity, but they may also be less dominant, surrounded by a diverse array of transportation options, including improved rail lines and bicycle networks.

Enormous investment opportunities could be created by an energy revolution of this magnitude. Just as the great fortunes of Rockefeller, Ford, and many others flowed from the turn-of-the-century oil boom, the next energy transition promises to create a new generation of successful entrepreneurs. Inevitably, however, there will also be losers. Today's petroleum refineries, oil tankers, nuclear power plants, and automobile factories could soon become stranded investments—many of them representing multibillion dollar losses once the environmental cleanup costs are accounted for. Major energy companies that fail to anticipate or plan for the coming energy transition are likely to face dim prospects. A recent analogy is the plight of IBM, which failed to anticipate the speed or direction of the personal computer revolution that it had helped create only a decade earlier—and soon found its bluechip financial status threatened.

How quickly the world's energy systems will change is one of the biggest uncertainties. Even the most conservative analysts might concede that the kind of energy systems sketched out here could evolve over the very long run. Our more accelerated timetable is based on a faith in human ingenuity, and on an expectation that government policies will evolve rapidly in the years ahead under the pressures of public opinion.

Of course, the enormous lobbies created by today's oil, coal, nuclear, and automobile industries will not be easy to overcome. In most countries, governments are more involved in the energy economy than in any other sector except the military, and their policies are heavily influenced by everything from the job aspirations of German coal miners to the research priorities of French nuclear scientists. The pace of the transformation will be determined in large measure by how quickly the new forces of change—both environmental and industrial—gain the upper hand. That day may be closer than many people expect. Despite failures to develop a number of highly touted energy solutions during the past two decades, the technological advances of that period have finally begun to reach critical mass. The forces of change are now gathering strength, as environmental movements flourish and begin to link up with emerging energy entrepreneurs.

History suggests that major energy transitions—from wood to coal or coal to oil—take time to gather momentum. But once economic and political resistance is overcome and the new technologies prove themselves, things can unfold rapidly. This is how today's energy systems emerged at the end of the last century, and it may be the way a sustainable energy economy begins to emerge at the end of this one. If so, the coming energy revolution will have profound effects on the way all of us work and live, and on the health of the global environment on which we depend.

2

Oil Shock

Only rarely does economic history have turning points as dramatic as the one on October 6, 1973. That afternoon, the eve of the Jewish holy day of Yom Kippur, hundreds of Egyptian and Syrian jets attacked Israeli positions near the Golan Heights, along the Suez Canal, and in the Sinai desert. The ensuing three-week air and ground war was one of the most intense and destructive conflicts the world had ever seen, including the loss of nearly 20,000 lives, 2,800 tanks, and 565 aircraft. Tensions grew as the United States used a massive airlift to resupply Israel, and by the final week of the war, the nuclear forces of the United States and the Soviet Union had gone to their highest alert status since the 1962 Cuban missile crisis.[1]

While the war itself was furious and brief, its deeper impact emerged gradually. Resentful of U.S. and Dutch

support for Israel in the war, the Arab nations announced an oil embargo against the United States and
the Netherlands on October 17th. Although it was never
all that firm and lasted only a few months, the embargo
came at a time when oil markets were already tightening
as a result of growing demand and limited supplies. The
result was a buying panic. By the end of the year, motor-
ists around the world were waiting in long gasoline lines,
and the price of crude oil had gone from less than $3 per
barrel to more than $13.[2]

The 1973–74 oil crisis marked an abrupt end to the
era of oil-fueled economic growth that characterized the
decades following World War II. Scores of global economic indicators—from gross national product (GNP)
growth to steel production—show sharp breaks around
1973. Among the lasting global effects were soaring in-
flation, slower economic growth, and high rates of
unemployment. In developing countries, which had
only recently come to rely on oil, the impact of higher
prices was particularly devastating, leading in many
cases to massive foreign debts and declining incomes.
Throughout the world, the easy assumptions of never-
ending economic and social progress began to be questioned.

For the global energy system, the 1973 oil shock and
the one that followed it six years later led to a particularly sharp change in direction. Throughout the fifties
and sixties, energy prices were generally falling, and energy use surged. Stable supplies of cheap oil were used
to fuel both industry and transportation, and global petroleum consumption was doubling every decade. High
and growing levels of energy use were at the time considered synonymous with twentieth-century economic success.

When all those assumptions were obliterated over-
night, energy planners were forced back to the drawing
board. The experts they turned to were equally divided
between those who retained the technological optimism
of the sixties, expecting either nuclear or solar energy
sources to ride to the rescue, and those who despaired
that anything could stave off an extended period of economic decline. In 1973, the U.S. government was at
work on a plan to shift from oil to coal, oil shale, and
nuclear breeder reactors fueled by plutonium.[3]

This proposal reflected the scientific confidence of the
time, for not even a prototype version of a commercial
breeder reactor was in operation anywhere in the world.
Two decades later, breeder reactors have only barely
made it to the prototype stage, and they have performed
so poorly that most nations have virtually abandoned
their development programs. In fact, nearly all the
"technical fixes" to the world's energy dilemmas that
were offered in the seventies—from orbiting solar satellites to synthetic fuels from coal—have fallen by the
wayside, leading many experts to doubt that any viable
alternatives to the current energy system can be found.

Although public attention tended to focus on the con-
voluted politics of energy supply that marked the late
seventies—particularly the contentious semiannual Organization of Petroleum-Exporting Countries meetings
that continue to be held, even in the nineties—hindsight
shows that the long-term impacts on the way energy is
harnessed and used are far more profound. Within a
decade of the oil embargo, evidence began to emerge
that changes in the world's energy systems were under
way. These developments were not only much simpler
than the ones envisioned by government scientists, they
also turned out to be more effective. And they laid the

groundwork for more profound changes in the years ahead.

*   *   *   *

Any effort to introduce new ways of harnessing or using energy runs headlong into the massive and seemingly impregnable global energy system that developed during the past eight decades. The outlines of that structure emerged in North America at the turn of the century. The Industrial Revolution was already well under way, but its energy requirements were still met mainly by solid fuels: wood, which had been used by society for millennia, and coal, a more abundant and concentrated fuel that caught on first in Great Britain and later in continental Europe and North America. This was a dirty and cumbersome energy system, but the lack of alternatives gave the solid fuels a near monopoly—supplemented by mechanical hydropower used to run mills and by small amounts of whale oil and synthetic coal gas that were used for lighting.

The petroleum era began slowly, almost imperceptibly, in the 1890s, as the first primitive "auto" mobiles (a term coined by the French) took to the road. Expensive and impractical at first, the new contraptions were confronted by many skeptics who doubted they could ever match the performance and reliability of horse-drawn carriages. In fact, it was initially unclear what kind of fuel or power system would prove most successful. A highly publicized race between Paris and Bordeaux in 1895 proved the advantages of internal combustion engines, and a gusher in Spindletop, Texas, in 1901 signalled the existence of a sufficient supply of petroleum to make oil a credible fuel.[4]

Subsequent advances in refining provided not only

ample supplies of gasoline but also the heavy, viscous hydrocarbons that turned out to have an important role—providing tires on which cars could run and asphalt pavement for the tires to travel on. The oil-powered transportation system essentially created its own infrastructure; as it gained economic and political momentum, oil began to shape society itself, opening up an era of nearly unlimited mobility and spurring profound social changes.

The final years of the nineteenth century witnessed the emergence of another new energy technology that proved equally central in shaping life in the twentieth. In New York, Thomas Edison started the world's first electric power company. In a Wall Street warehouse, Edison connected a coal-fired boiler to a steam engine and dynamo, then linked the plant by underground wire to a block of nearby office buildings. When the switches were finally flipped at the Pearl Street Station on September 6, 1882, 158 light bulbs (also designed by Edison) flashed on, and the Edison Electric Illuminating Company made converts of its carefully chosen first customers—J.P. Morgan and the *New York Times*.[5]

Electric power was an immediate hit—turning its pioneer into a millionaire and causing the stock of competing gas companies to plummet. Edison viewed electricity as a dynamic, competitive service business; initially he even sold lighting to his customers by the bulb. In the decades that followed, electric power companies quickly proliferated, offering both direct and alternating current at various voltages, and often running competing electric lines down opposite sides of the same street.

A century later, Edison would hardly recognize the electric power industry—nor its ubiquitous effects on societies around the globe. With all the world's cities and many of its villages now wired, electric current flows from power plants thousands of times larger than Edison's into millions of homes, providing power for "necessities" such as refrigerators, television sets, and computers. Electric power has not only transformed life and work, it has fostered a global enterprise with annual revenues estimated at more than $800 billion—roughly twice the size of the world auto industry.[6]

The twin engines of the petroleum and electric power industries have in many ways shaped the twentieth century, unleashing everything from the destructive power of World War II to the suburbs of the fifties and the computer age that followed. The fifties and sixties—the golden era of cheap energy—offered little reason to question the wisdom of pursuing this path. They were a time of enormous optimism, and few industrial leaders or government policymakers saw anything worrisome in the exponential growth trends. Few bothered to notice the surprisingly narrow technological path—and heavy oil dependence—on which all this was based.

Starting first in North America, but spreading quickly to Europe and Japan, economic growth soared, and with it, demand for energy. Direct use of solid fuels such as coal and wood was largely phased out in most industrial countries between 1950 and 1970 in favor of oil, electricity, and natural gas. From Oslo to Osaka, electrically lit and air-conditioned skyscrapers became commonplace, while reliance on oil-intensive auto, truck, and air transportation also took hold. By the early seventies, more than three times the amount of energy was being used worldwide than in 1950, while the use of oil had risen more than fivefold. (See Figure 2–1.)[7]

The postwar energy boom would not have lasted long without the exploitation of the world's largest oil fields,

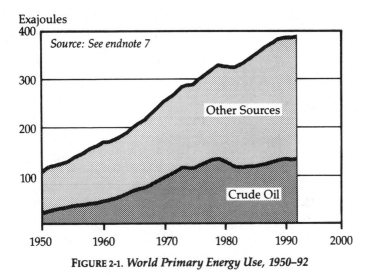

FIGURE 2-1. *World Primary Energy Use, 1950–92*

in the Middle East. Earlier periods of energy growth had required relatively minor trade in fuels, but during the fifties and sixties, countries that had virtually no oil of their own showed little hesitancy about shifting to an oil-based economy. Japan, for example, made a deliberate decision to phase out its high-cost domestic coal production in the sixties, and many developing countries substituted oil products for traditional wood fuel. And even the United States, which had led the world into the age of oil, saw its production peak in 1972 and its need for Middle Eastern oil balloon. By then, hundreds of oil tankers had fanned out from the Middle East to supply much of the world with petroleum. Oil had become not only the dominant energy source but a sort of safety valve that could be easily and cheaply substituted for coal in Japan, fuelwood in India, and other energy sources elsewhere.[8]

Few realized in the early seventies just how quickly the global oil market was tightening. With the world economy expanding at 4 percent a year, oil use was doubling every decade. The Arab countries' market power was growing rapidly, but even these nations would only be able to sustain such growth for so long. At the time, a few analysts began to warn that some sort of crisis might lie ahead, though even they thought that it was probably 5–10 years away. As is often the case when distant clouds appear on an otherwise sunny horizon, these sporadic warnings were generally dismissed as inconsistent with the bullish spirit of the times.[9]

The 1973 oil embargo turned out to be only the first energy shock of the decade. In January 1979, the Shah of Iran fled Tehran, culminating a year-long struggle by both secular and religious leaders who resented the royal family's corrupt and despotic regime. As part of the sometimes violent campaign that sent the Shah packing, strikes by Iran's oil fields workers had cut off exports from what was then the world's second largest exporter. The resulting chaos and panic in world markets led to a near tripling in world oil prices, spurring an even deeper recession than the one earlier in the decade. It was not until 1986 that prices fell to less than $20 per barrel—still 50 percent above the average price that prevailed during the previous century. (See Figure 2–2.)[10]

According to the wisdom of the seventies, only the quick development of a magical new fuel or technology could get the world economy back on the growth track. The result was a series of government-funded crash efforts to commercialize everything from breeder reactors to solar power plants and coal-based liquid fuels. The plans reflected a common faith that direct use of electricity—a versatile, "modern" energy carrier—was the

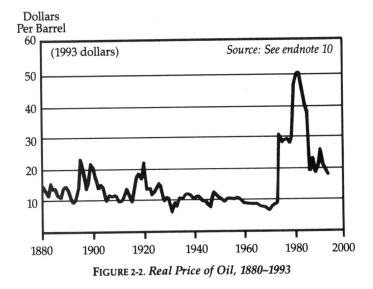

Dollars
Per Barrel

FIGURE 2-2. *Real Price of Oil, 1880–1993*

best alternative to oil. At a time when U.S. astronauts had recently reached the moon and some scientists were predicting an imminent cure for cancer, optimism about energy technologies did not seem out of place. At legislative hearings and cocktail parties, the phrase "if we can go to the moon, surely we can. . ." was often heard. Using that philosophy as a guide, many billions of dollars in government funds have been invested in a range of new energy technologies over two decades—often with disappointing results.

Nuclear power was probably the greatest disappointment, confounding the nearly unanimous scientific consensus that had so enthralled government leaders in 1973. Despite huge subsidies and aggressive public relations efforts, the construction of new reactors began to decline soon after the first oil crisis—slowed by rising

costs and growing public concern that was later heightened by the accidents at Three Mile Island in 1979 and
at Chernobyl in 1986. Starting in the United States in
the late seventies, then spreading to Western Europe in
the mid-eighties and the former Eastern Bloc by the
nineties, most nuclear expansion programs were
halted.[11]

By 1990, after two decades of rapid growth, global
nuclear output had levelled off (see Figure 2–3)—supplying at the time 17 percent of the world's electricity,
and 5 percent of its total primary energy. It now appears
that there will be little if any growth in the use of nuclear
power between 1990 and 2000, and many plants built in
the sixties and seventies may soon be retired. Even the
strongest proponents of nuclear energy are left to argue
that if governments went back to the drawing board and

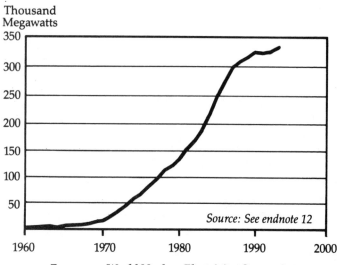

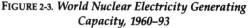

FIGURE 2-3. *World Nuclear Electricity Generating
Capacity, 1960–93*

designed a simpler, safer kind of reactor, nuclear orders might revive—but not for a couple of decades. Never before has a major new technology regressed so rapidly.[12]

As the nuclear star faded in the eighties, coal became the alternative of choice for many nations trying to reduce their dependence on petroleum. Worldwide, its use rose more than 30 percent between the mid-seventies and the late eighties. A few countries even followed the path of Japan: importing coal in order to replace imported oil. But the eighties coal boom was confined mainly to electric power generation and did not displace the much larger uses of oil in transportation and industry. As a bulky, dirty, solid fuel, coal was simply not a ready substitute for petroleum. Several governments' efforts to encourage the conversion of coal to synthetic liquid fuels quickly ran aground—beset by unexpected technical glitches and huge cost overruns. By the end of the decade, nearly all these projects had been abandoned.[13]

In the world's most populous countries, China and India, coal plays a much larger—and still growing—role, however. Both have huge coal reserves, and neither has sufficient funds to pay for much imported oil. As a result, coal is used extensively for home heating and cooking as well as to fuel factories, run railroads, and generate electricity. In China, coal accounted for 76 percent of the country's commercial energy supply in 1990, a situation much like Great Britain's in the late nineteenth century.[14]

Oil was replaced by other energy sources as well. Reliance on hydropower grew steadily, particularly in developing countries. Many lumber mills and paper plants in Canada and the United States increased their burning of

scrap wood, and geothermal energy use expanded in the
Philippines. The use of natural gas rose rapidly in some
countries, particularly the Soviet Union, but it fell in
others where supplies temporarily ran short. (See Chapter 5.) Altogether, oil's share of world energy use fell
from 36 percent in 1970 to 31 percent in 1992. (See
Table 2–1.)[15]

In the United States, where comprehensive figures are
available from 1850 onward, the evolving shape of the
world energy economy can be seen in microcosm: Oil is
still number one, at 37 percent, but its share is falling,
while coal and natural gas are each in the mid-twenties,
with coal plateauing and gas rising. Other energy

TABLE 2-1. *World Primary Energy Use by Source,*
1970 and 1992

Source	1970		1992	
	Use	Share	Use	Share
	(exajoules)	(percent)	(exajoules)	(percent)
Crude Oil[1]	92	36	123	31
Coal	65	26	91	23
Natural Gas[1]	42	17	82	21
Biomass[2]	39	15	50	13
Hydropower[3]	13	5	24	6
Nuclear[3]	1	0.3	22	6
Geothermal, Wind, Solar[3]	< 0.1	< 0.05	0.4	0.1
Total[4]	252	100	392	100

[1]Natural gas liquids are included under natural gas, not crude
oil.   [2]Counts energy that may be consumed in boiling off water
content when wet fuels are used.   [3]Converted on the basis of the
average thermal efficiency of a modern fossil fuel steam plant (33
percent).   [4]Columns may not add up to totals due to rounding.

SOURCE: See endnote 15.

sources—nuclear power, hydropower, and biomass—each contribute less than 8 percent of the total energy supply, and are essentially static. (See Figure 2–4.) Solar and wind energy have yet to reach the 1-percent share that would give them a place on U.S. or global energy charts.[16]

While recent supply trends are somewhat ambiguous, other developments are clearer—particularly the sharp reversal in consumption trends. World oil use declined more than 10 percent between 1979 and 1985, then rebounded in the following years. Still, even in 1993 world oil use was only 12 percent above the 1973 level. (By comparison, world population grew more than 40 percent during the same period, and economic activity, 65 percent.)[17]

This decline in oil dependence resulted largely from conservation efforts, particularly those aimed at improv-

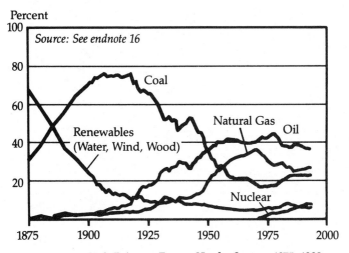

FIGURE 2-4. *U.S. Primary Energy Use by Source, 1875–1993*

ing the efficiency with which oil is used. (See Chapter
4.) The trend is most obvious in U.S. automobiles: after
decades of stagnant fuel economy, average new-car efficiency doubled from 14 miles per gallon in 1974 to 28 in
1985 (going from 16.8 liters per 100 kilometers to 8.4
liters). Reduced weight, smaller engines, better aerodynamics, improved transmissions, and smoother tires
were among the relatively simple improvements made
by automakers. Many industrial facilities were able to
cut their use of energy by a similar order of magnitude.
In North America, Europe, and Japan, the best overall
measure of energy productivity—the amount of GNP
produced per unit of energy used—has risen by 40–45
percent since the early seventies, slashing the amount of
oil and other fuels needed to run the economy.[18]

Equally important was the impact that 12 years of
high oil prices had on the developing world. Although
the enormous transfer of wealth to the Middle East during the eighties came largely at the expense of industrial
countries, poor nations suffered the most lasting damage. In 1973, many of these countries were poised to
adopt oil- and electricity-based energy systems—plans
that were quickly cut short. Most found themselves
spending much of their hard-currency reserves on imported oil denominated in U.S. dollars throughout the
seventies and eighties.

Three fourths of developing countries were oil im-
porters in 1987, and of the 38 poorest countries, 29 had
to bring in more than 70 percent of their commercial
energy—nearly all in the form of oil. Even at today's
lower oil prices, most sub-Saharan African countries
spend a quarter to half of their hard-currency earnings
on petroleum imports, effectively gutting investment in
other areas.[19]

This drain contributed to the world debt crisis of the eighties and to the subsequent decline in per capita incomes in much of Africa and Latin America. During the past decade, India's oil import bill totalled $36.8 billion—equal to almost 87 percent of its new debt and soaking up nearly a third of its export receipts.[20]

One consequence of this economic slide was that growth in energy use slowed in developing countries. By 1992, the Third World—with nearly 80 percent of the world's population—still used only 35 percent of the world's primary energy. (See Table 2–2.) Unlike the situation in industrial countries, this resulted less from im-

TABLE 2-2. *World Primary Energy Use by Region, 1970 and 1992*

Region	1970		1992	
	Energy Use	Per Capita	Energy Use	Per Capita
	(exa- joules)	(gigajoules per person)	(exa- joules)	(gigajoules per person)
Industrial	182	174	253	207
North America	77	338	99	351
Western Europe[1]	48	136	63	167
Central Europe	43	124	67	161
Indus. Australasia	14	120	23	160
Developing	70	26	139	33
China	22	26	39	33
Developing Asia	22	21	47	27
Africa & Mid. East	14	30	31	36
Latin America	12	41	22	49
World[2]	252	68	392	72

[1]Includes former East Germany. [2]Columns may not add up to totals due to rounding.

SOURCE: See endnote 21.

proved efficiency than from a general slowdown in economic growth.[21]

★ ★ ★ ★

Most analysts agree that the collapse of oil prices in 1986—when Saudi Arabia decided to relax controls on production and reclaim its share of the market—represented a turning point nearly as important as the one in 1973. The price of oil reached a brief low of less than $10 per barrel in 1986, and has since ranged between $15 and $25, less than half the peak level recorded in the early eighties. In Japan, the cost of oil denominated in yen is actually less than it was before the oil embargo (partly due to the strength of the yen), while in the United States it costs less on average today to drive a kilometer, thanks in part to the doubling of fuel economy.[22]

Falling oil prices naturally spurred a period of rapid economic growth and an increase in oil use. So far, the oil market has been able to absorb the resulting increase in demand, with prices holding relatively steady except in late 1990 during the buildup to the Gulf War. In most countries, the "energy security" issue that so dominated politics in the early eighties has faded. Yet midway through the nineties, a lively debate has reemerged in government and academic circles about the dangers of depending too heavily on Middle Eastern oil, and whether another petroleum crisis may be around the corner. The answer to this question will affect the pace of change in world energy trends in the years ahead.

Already, the recent period of low oil prices has begun to undermine the conditions that created it. For one thing, automobile fuel economy is barely increasing in most countries, and the use of cars is soaring in the developing world. The most dramatic change is apparent in Asia, where oil consumption has increased 50 percent since 1985 and is projected to continue rising. (In South Korea, it rose 21 percent in 1992 alone.) Similar binges are under way in Latin America and the Middle East. Oil use is growing much more slowly in the industrial regions of North America and Europe, but even there, further growth is expected by most forecasters. Meanwhile, oil use actually declined by 25 percent in the rapidly weakening economies of Eastern Europe and the former Soviet Union during the early nineties, temporarily masking the increases elsewhere. Most forecasters expect a spurt in world oil demand in the late nineties.[23]

Oil outside the Middle East remains limited. Most of the large fields tapped over the past two decades in Mexico, Alaska, and Siberia are no longer expanding, and in some cases they are shrinking. In the United States, now the world's number two producer, daily output has decreased from nearly 9 million barrels in 1985 to less than 7 million barrels in 1993, pushing U.S. imports from 3.2 million barrels a day to 6.7 million—accounting for a record 49 percent of U.S. oil consumption. Government analysts expect further declines of at least a million barrels a day in U.S. production by the end of the decade. As a result of dwindling domestic prospects, major U.S. oil companies gradually have shifted more than half their exploration investments from Texas, Oklahoma, and other domestic locations to higher-yielding overseas prospects.[24]

The continuing decline in U.S. production may be partly offset by minor increases in Asia and Latin America, but this would still leave total production outside the Middle East and the former Soviet Union flat during the rest of the decade. The most striking shift in the oil market in the early nineties, however, is the collapse of what was the world's number one producer: by early 1994, daily oil production in Russia had fallen to less than 7 million barrels, down from the peak of 11.5 million barrels in 1988. The drop can be traced to a series of problems—aging equipment, shortages of spare parts, poor management, and, in some cases, rapidly depleting oil fields. Since oil demand in the former Eastern bloc has fallen almost as dramatically, the short-run effect on the world market has been limited. Still, even if Russian exports fall no farther and those of Azerbaijan, Kazakhstan, and Turkmenistan rise slightly, the former Eastern bloc as a whole is likely to put more pressure on world oil markets in the future, not less.[25]

Nevertheless, as of 1994 the world oil market was closer to glut than shortage, with 3–4 million barrels per day of unused production capacity putting downward pressure on prices. And future crises, some analysts now argue, are likely to be shorter-lived and less disruptive than those of the seventies. Philip Verleger of the International Institute of Economics, for example, believes that the more competitive, flexible, less vertically integrated oil market of the nineties—including large "spot" and "futures" markets—will allow the world to respond more effectively to future crises.[26]

Despite these encouraging developments, the evidence still suggests that the world oil market is entering a danger period. Much of the current excess capacity represents unused Iraqi production—compelled by a U.N. embargo that began shortly before the Gulf War— and virtually all the rest lies in the Middle East as well. This leaves the world facing the same central oil problem that emerged two decades ago: two thirds of the world's remaining proven reserves are located in a narrow crescent running from Iran in the north to the United Arab Emirates in the south.[27]

This region has been the world's marginal oil supplier for more than 20 years, and its share of the global market is now reaching new highs. (See Figure 2–5.) By 2000, a growing world economy is projected to need 5–8 million additional barrels of Middle Eastern oil each day, boosting the region's share of the world oil market from nearly 31 percent in 1993 to close to 40 percent by the end of this decade.[28]

The prospects for meeting this demand are uncertain. Some 2 million barrels a day will be available once Iraq returns fully to the world market, and several other Persian Gulf countries could increase production somewhat. But even the abundant oil fields of the Middle East are beginning to show their age—and their limits. Production costs are rising, and future increases in ca-

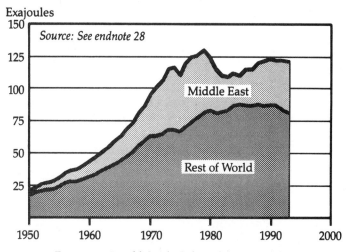

FIGURE 2-5. *World Crude Oil Production, 1950–93*

pacity are likely to be modest, expensive, and relatively slow.

Saudi Arabia, which has the greatest potential to raise output, is now spending $2 billion a year just to maintain its current daily capacity of 10 million barrels. A 1993 study by the Center for Global Energy Studies, headed by former Saudi oil minister Sheik Yamani, suggests that Saudi Arabia may never supply the 12, 15, or 18 million barrels of oil a day that many western politicians and industrialists seem to be counting on.[29]

If an all-out effort to increase Middle Eastern production is needed just to keep up with demand, this leaves little room for unexpected political turmoil. Saudi Arabia, the traditional "swing" producer, may soon have the production accelerator all the way to the floor—whether it wants to or not. The last time that was needed, in 1979, a modest disruption in Iran's oil flow resulted in a near tripling of prices. The health of the world oil market therefore will continue to depend on the stability of a handful of Middle Eastern regimes, particularly that of Saudi Arabia, whose current 20-percent share of total world exports will almost certainly rise in the years ahead.[30]

The Persian Gulf region has been a tinderbox for decades, but now it faces greater economic and social stresses than ever before. The nations of the region are characterized by large wealth disparities, high rates of population growth, autocratic political systems, and rapid social change. Most immediately, these countries face serious financial stresses generated by the recent period of low oil prices and the economic drain of the 1991 Gulf War. Saudi Arabia has already dissipated the better part of $120 billion in assets that it earned during the price run-up to the war, and is projected to owe

$100 billion by the end of 1994. Even with the possible settlement of the Israeli-Palestinian conflict and a general rapprochement between Israel and the Arab states, the Middle East is likely to remain dangerously unstable.[31]

Although the world oil market has lost its ability to command political attention or shape energy trends for the time being, oil could easily reemerge as a powerful economic force within the next decade. Whether this occurs depends on a series of political variables that are impossible to predict. Regardless of what happens with oil, however, a range of other forces—principally environmental—are now laying the groundwork for an energy transition.

3

Eco Shock

The end of the Gulf War in early 1991 marked another turning point for the world energy system. What had begun as a familiar Middle Eastern oil crisis, with clashing armies and soaring fuel prices, ended with another kind of debacle. Just before they retreated, Iraqi troops released 3 million barrels of oil into the Persian Gulf and even more into the desert, and then torched more than 500 oil wells. Kuwait was transformed into a scene out of Dante's inferno, with pillars of flame lighting dark pools of petroleum, oily smoke choking the capital city, and heavy ash falling from the sky hundreds of kilometers downwind in Iran. For close to a year, the Kuwaiti oil fields burned, at their peak consuming 6 million barrels of oil a day and releasing 10 times as much pollution as all U.S. power and industrial plants combined.[1]

The Gulf War's quick transition from energy and military crisis to ecological calamity is symbolic of a wider shift in the last decade. By the early nineties, energy markets were being shaped more by environmental problems and responses than by the economics and politics of oil. First, the *waldsterben* (forest death) crisis that struck central Europe beginning in 1982, caused mainly by coal-burning power plants, provoked a tightening of air pollution regulations. Then the explosion at the Chernobyl nuclear plant in Ukraine in 1986 sent a radioactive plume across much of central and western Europe, causing a rapid meltdown of the already declining world nuclear industry. In 1989, the wreck of the Exxon Valdez left an extensive oil slick in the ecologically fragile Prince William Sound, reminding the world of one of the continuing prices of using 60 million barrels of oil a day.[2]

Finally, starting with a serious heat wave and drought in North America and fed by prominent statements and reports from scientists, the world came face-to-face with what may turn out to be the ultimate limit to the fossil fuel economy: the capacity of the atmosphere to absorb ever growing quantities of heat-trapping carbon dioxide gas. Recognizing the risk of rapid climate change, in June 1992 the 106 heads of state or government at the Earth Summit in Rio de Janeiro signed a treaty designed to stabilize the earth's climate.[3]

Although many of the environmental costs of our twentieth-century energy economy have been known for a long time, by the early nineties even the energy traditionalists—from consultant and author Daniel Yergin to the bureaucrats at the International Energy Agency—had proclaimed "environmentalism" a potent force in world energy markets. Environmental groups that had largely abandoned their energy work in the early eighties made it one of their main priorities by the end of the decade, encouraged by the concerns of their members and the support of foundations.[4]

In many circles, the "energy security" mantra of the seventies and eighties is being replaced by the principle of "sustainable" energy, now used in the crafting of long-range scenarios and strategic plans. Royal Dutch Shell, one of the world's largest oil companies, captured attention in 1991 by publishing a "Sustainable World" scenario that pointed to the possibility of severe limits on emissions of carbon dioxide and other pollutants. In Washington, the Business Council for a Sustainable Energy Future was formed in 1992, bringing together a broad coalition from the natural gas, electric power, renewable energy, and energy efficiency industries to advocate reform of government energy policies.[5]

Yet in most nations, such concerns run head-on into the powerful forces that protect today's $2-trillion world energy system. The large oil, coal, nuclear, and automotive industries—and many of their associated unions—continue to fight effectively for the government subsidies they have jealously guarded over the years. Whether it be coal in Germany, oil in Nigeria, or nuclear power in France, it is clear that many industries still receive support that directly undermines the goal of a "sustainable" energy system.[6]

In the face of growing evidence that this panoply of environmental problems can only be addressed through systemic changes in the energy economy, conflicts between new environmental priorities and the burdens of past interests are likely to grow. The outcome of these battles will help determine the pace of change.

*   *   *   *

That heavy use of fossil fuels can damage human health and the natural environment is hardly a recent discovery. As early as 1306, King Edward I of England banned the burning of coal in London in order to reduce the heavy air pollution already choking the city. Despite such efforts, coal use skyrocketed during the Industrial Revolution, so that by the late nineteenth century, London was known for its unbreathable air and blackened buildings. As late as 1952, thousands of Londoners died when stagnant air trapped city residents in their own waste. Although this tragedy resulted in a tightening of British air laws, the air quality crisis was only just beginning, for the use of fossil fuels—particularly oil—was about to go through the roof.[7]

The full impact of the age of oil first became clear in Southern California, where postwar economic growth and immigration coincided with the arrival of affordable automobiles and government-subsidized "freeways." By accident of geography, Los Angeles sits in a basin that allows stagnant air masses to create "inversions," in which a layer of warm air can trap pollution near the surface for days or weeks, while the sun triggers chemical reactions that create more pollutants. Starting in the late forties, the Southern California air pollution problem mushroomed, and with it, public concern. By 1953, local governments had passed their first air pollution laws, which were shortly followed by state-level standards, and then the U.S. Clean Air Act of 1970.[8]

During the seventies and eighties, virtually all industrial and many developing countries enacted air quality standards and emissions requirements. California and

Japan led the way, with Europe trailing behind, but following the acid rain crisis of the eighties, tough national and Community-wide pollution standards were enacted, in some cases surpassing those in the United States.

Most laws focused first on reducing a few obvious pollutants—principally hydrocarbons, carbon monoxide, and particulates. Among the most obvious results were modest changes in automobile engines, the addition of catalytic converters to the tailpipes of cars, and the installation of electrostatic precipitators to control the ash emitted by power plants. More recently, flue gas desulfurization units have been added to many new power stations. Although the first generation of pollution controls were not that effective and deteriorated with use, they gradually improved—allowing a 90- to 95-percent reduction in key pollutants from new cars by the late eighties, and a 95-percent reduction in the sulfur dioxide emitted by new power plants.[9]

Rapidly growing energy use has unfortunately obviated some of the gains. Total U.S. emissions of sulfur dioxide had only fallen 26 percent from their peak by the mid-eighties, while nitrogen oxides were down 15 percent. Since then, each has levelled off. (See Figure 3–1.) In 1991, the U.S. Environmental Protection Agency estimated that 86 million Americans lived in counties that still did not meet federal ozone standards. In southern California, for example, peak ozone levels have fallen 55 percent since the mid-fifties, but are still higher than anywhere else in the United States, forcing government agencies to look for new strategies.[10]

Compared with many parts of the world, however, California's air is pristine. An estimated 1.3 billion people worldwide—most of them in developing countries—

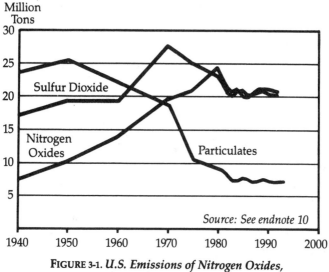

FIGURE 3-1. *U.S. Emissions of Nitrogen Oxides, Sulfur Dioxide, and Particulates, 1940–92*

now live in areas that do not meet World Health Organization standards for atmospheric particulates, causing between 300,000 and 700,000 premature deaths each year. Similarly, more than 1 billion people in urban areas are exposed to unacceptable sulfur dioxide pollution. The rapid growth of heavy industry and a near quintupling in the number of automobiles outside western industrial countries—from about 22 million in 1970 to more than 109 million in 1991—have caused steady deterioration in the air quality of virtually all Third World cities.[11]

Even in China, where the bicycle is still the leading form of transportation, Toyotas and Mercedes are becoming a common sight on city streets, exacerbating the already miserable air quality. During the next decade, growing urban populations and rising incomes, particularly in Asia, are likely to increase the number of people who can afford a car or a motorbike, leading to even more pollution.

Mexico City is widely thought to have the world's dirtiest air, with concentrations of carbon monoxide, ozone, and particulates that exceed legal limits nearly every day it does not rain. According to the government, the city's residents enjoyed only 31 days with air considered safe to breathe in 1993. Similar horror stories are repeated regularly in cities ranging from Athens to Cairo and Jakarta. Breathing the air in Bombay is now the equivalent of smoking 10 cigarettes a day. In Bangkok, which has 2 million automobiles and thousands of factories, a million people were treated for respiratory problems in 1990, lead poisoning is lowering the measured intelligence of children, and lung cancer is three times as common as in the rest of the country.[12]

In China, the world's leading coal consumer—much of it high in sulfur—pollution control equipment ranges from inadequate to nonexistent, and air quality is fast becoming an ecological crisis. Extensive air-pollution-related forest damage has been identified in some areas where acidity levels are building in the soils, and the level of suspended particles in Chinese cities is 14 times that in the United States. Medical researchers report growing numbers of eye irritations, lung disorders, and heart ailments. According to an extrapolation from U.S. data by Keith Florig of Resources for the Future, more than 900,000 people in China may be dying annually as a result of pollution-related lung disease, a number that is likely to increase as the cumulative effect of decades of polluted air is felt.[13]

As poor countries begin to catch up with the levels of fossil fuel use found in richer nations, many people living there are demanding change. Their governments are beginning to adopt tougher air quality standards. Some require that emission controls be placed on new power plants and catalytic converters on cars, as well as mandating the use of cleaner, lead-free gasoline. Since the required technologies have already been commercialized and are much less expensive than in the eighties, the economic argument against environmental progress in developing countries has weakened.

One obstacle still plaguing rich and poor nations alike is that air pollution has turned out to be more complex than once assumed, involving veritable soups of chemicals that react in unpredictable ways. Scientists have discovered, for example, that nitrogen oxides are often the key factor in the formation of ozone, a lung and throat irritant. In response, some of the air laws passed since the late eighties require the introduction of reformulated gasoline, the use of "three-way" catalytic converters for cars, and new catalytic reduction devices that cut nitrogen oxide emissions at power plants. Together, this wave of new laws is likely to enhance the air quality in many countries, but will also add to the cost of using fossil fuels.[14]

Another problem that has loomed larger in recent years is the emission of highly toxic chemicals, including heavy metals such as lead, mercury, cadmium, and arsenic. Such substances, released into the air by power plants, oil refineries, and other facilities, often enter food and water supplies. The long-term health consequences of this are still unclear. Although the lead problem is being addressed by reducing the use of lead additives in gasoline in many countries, the others are more intractable. Coal and oil both contain trace amounts of these metals, which would be expensive or impossible to remove. In addition, the ash collected from coal-burning power plants often contains heavy metals that leach into nearby streams. And most large oil refineries have stored thousands of tons of toxic chemicals over the decades. Indeed, under current law, many of these refineries will be classified as hazardous waste sites once they close.[15]

The fossil fuel economy has additional costs connected with the extraction of fuels from the earth. Although their safety record has improved, coal mines still take a high toll in injury and death to coal miners, mainly as a result of mine accidents and black lung disease. As the coal industry has shifted to open-pit or strip mining in recent decades, part of the human toll has been replaced by an ecological one. Huge mechanical shovels and drag lines scrape off the surface soil and rock in order to extract the coal beneath. The resulting gashes in the earth are often more than 100 meters deep and can cover tens of square kilometers. Although limited restoration of such sites is required in some countries, strip mining is still largely uncontrolled in others. The massive quantities of rock removed often contain toxic substances that leach into nearby streams and groundwater. Around some mines, the rivers are biologically dead and the water undrinkable, while others have been linked to sickness in nearby communities.[16]

The human toll of strip mines can be tremendous. At Singrauli, 150 kilometers south of India's holiest city, Varanasi, more than 200,000 people have been displaced by 12 open-pit coal mines and related power plants. Many of them, who were mainly farmers before the coal development started some 30 years ago, have been forced to move two or three times without compensation and can barely eke out a living. Even today, some live inside the coal mines. The mining operations and the disruption to the local environment caused by them have made tuberculosis and malaria endemic in the area.[17]

★   ★   ★   ★

In the history of the world energy economy, June 23, 1988, may turn out to be as significant as the day the Arab-Israeli war began. On that stiflingly hot day in Washington, D.C., a U.S. Senate committee heard stark testimony from James Hansen, the director of NASA's Goddard Institute for Space Studies and a leading government atmospheric scientist. Rapid climate change was likely in the decades ahead as a result of human activities, Hansen stated. To bolster his case, he presented the results of the latest computer models as well as a series of global average temperatures starting in 1880 that show significant warming, particularly in recent decades. The eighties, Hansen pointed out, was the warmest decade ever recorded.[18]

Only rarely are public policy turning points so clearly marked. Although much of the key data had been developed and even published in preceding years, Hansen's testimony was more definitive. A sober government scientist was publicly stating his conclusion that human-induced greenhouse warming is a prospect societies no longer can afford to ignore. Moreover, the environmental threat he was describing was unlike any other: long-term and essentially irreversible in nature.

Against the background of a persistent North American heat wave and drought, shock waves emanated quickly from Capitol Hill. As a result of enhanced public attention, research on climate change was rapidly increased in the early nineties—reaching $1.4 billion in

the United States alone in 1994. At the same time, dip-
lomats and scientists began a series of international
meetings aimed at achieving a treaty to protect the atmosphere.[19]

The concern of Hansen and other scientists is rooted
in the fact that the earth's climate is a delicate balance of
solar energy inputs, chemical processes, and physical
phenomena. Some gases, such as water vapor, carbon
dioxide, and methane, tend to absorb heat in the same
way that glass traps heat in a greenhouse, allowing temperatures to build. Although carbon dioxide now makes
up less than 0.04 percent of the earth's atmosphere, it
has played a crucial regulating role, rising and falling in a
pattern that is closely correlated with global temperatures.[20]

The notion that human activities might disrupt this
carbon dioxide thermostat was first proposed by Swedish chemist Svante Arrhenius in 1896. Coal and other
carbon-based fuels such as oil and natural gas release
carbon dioxide during their combustion. Arrhenius
theorized that the rapid increase in the use of coal in
Europe since the Industrial Revolution would cause a
gradual rise in global temperatures.[21]

Measurements taken from air bubbles trapped in gla-
cial ice suggest that the atmospheric carbon dioxide
concentration prior to the Industrial Revolution was
about 280 parts per million. But at a measuring station
in Hawaii, scientists have registered a steady increase
since then—from 315 parts per million in 1958 to 357 in
1993. (See Figure 3–2.) Thus the 13-percent increase in
carbon dioxide concentration in the last 35 years ex-
ceeds the rise in the previous two centuries, bringing the
concentration to its highest level in more than 160,000
years.[22]

Since the sixties, satellite reconnaissance, better data on past climate trends, and improved understanding of the oceans led to advances in atmospheric science, though it remained a relatively neglected field. The advent of powerful "super" computers provided an important breakthrough, allowing scientists to build models that simulate atmospheric effects, which permitted the first projections of how rapidly the climate might change.[23]

During the eighties, scientists detected increases in other greenhouse gases, notably chlorofluorocarbons (CFCs) and their substitutes (HCFCs and HFCs), methane, and nitrous oxide, which emanate from many sources—swamps, landfills, rice paddies, and industrial processes. Although each of these gases exists in the atmosphere in far smaller quantities than carbon dioxide does, together they have roughly one third as much greenhouse potential, in effect increasing the rate of projected climate change by 50 percent. At the same time, emissions of carbon have continued to increase, albeit at a slower rate since the 1973 oil crisis, reaching 6 billion tons annually in the early nineties. (See Figure 3–3.)[24]

To assess the magnitude of the problem and assist policymakers, the United Nations formed the Intergovernmental Panel on Climate Change (IPCC) in 1988, with more than 150 leading climate scientists from around the world. After two years of meetings and assessments, the IPCC concluded that although there is still much to be learned about climate change, the global circulation models have demonstrated their ability to simulate observed climate phenomena (confirmed again following the impact of the 1991 eruption of Mount Pinatubo) and are sufficiently reliable to provide a warning of future trends.[25]

Parts Per Million
(by Volume)

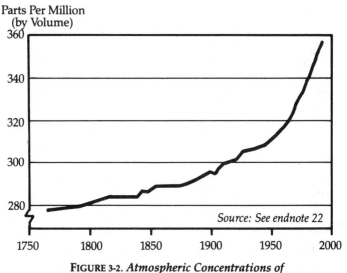

FIGURE 3-2. *Atmospheric Concentrations of*
Carbon Dioxide, 1764–1993

Billion
Tons

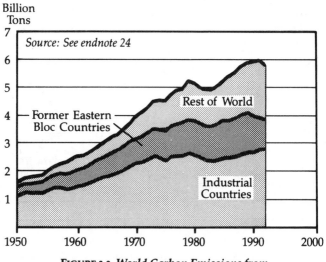

FIGURE 3-3. *World Carbon Emissions from*
Fossil Fuel Burning, 1950–92

In its reports, the IPCC predicted a rapid rise in global temperatures of 1.5–4.5 degrees Celsius by late in the next century. (See Figure 3–4.) It further stated that such an increase, particularly if it is near the upper end of that spectrum, could cause such sweeping changes in patterns of rainfall, drought, and temperatures that it would be highly disruptive to world agriculture and to many natural systems, including forests. Economists using these figures have estimated possible economic losses in the tens of billions of dollars annually.[26]

In 1992, the IPCC reiterated its earlier findings with a subsequent report, this time with the help of nearly 500 scientists. But even after this more recent report, debate has continued between supporters and detractors of the consensus it represents, most of it played out in the popular press rather than scientific journals. Critics such as

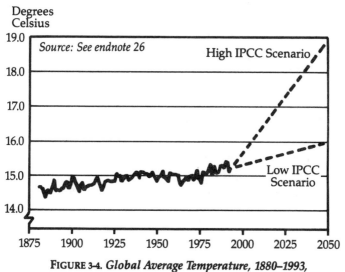

FIGURE 3-4. *Global Average Temperature, 1880–1993, With Projections to 2050*

Patrick Michaels of the University of Virginia and Richard Lindzen of the Massachusetts Institute of Technology argue that the climate models are too flimsy to be predictive, that the climate record shows a slower rate of warming than the models suggest, that increasing cloud cover may mitigate the effects of additional greenhouse gases, and that even if the climate were to change, the effects would be manageable and perhaps even beneficial.[27]

Although these scientists have seized effectively on the uncertainties inherent in computer-generated climate models, they have failed to produce a solid alternative that is viewed as credible by others. Indeed, the IPCC scientists acknowledge the uncertainties in today's computer models, and their report includes a broad range of scenarios (some of which would be even worse than the mid-range case that is generally described in the press). This lack of certainty is hardly a reason for complacency, however. The nature of the climate problem—with its long time lags, irreversibility over hundreds of years, and potentially catastrophic effects—argues for caution.

Many scientists believe that by the time they can predict future changes in detail, it may well be too late to avoid the consequences. Given that we are now conducting an unregulated experiment with the earth's only atmosphere, the burden of proof should lie with those who believe there is no problem. Although understanding of climate change is likely to grow in the years ahead, this issue is unlikely ever to be marked by the kind of precision that would allow societies to know precisely how bad things will get—or exactly how much carbon emissions need to be reduced. Rather, the focus will continue to be on questions of risk and uncertainty:

whether to reduce the chances of an indefinite but potentially devastating future.

Although this may seem like a fuzzy basis on which to make policy or invest billions of dollars, there is one industry in which such an exercise is routine: the insurance business. Insurance companies use risk assessments to decide which items to cover and how large a premium to charge. It is not surprising therefore that this industry, one of the world's largest, is one of the first to take climate change seriously. Alarmed by an unprecedented series of catastrophic wind storms and floods that caused billions of dollars in insured losses in the early nineties, insurance companies have begun to evaluate the prospect of climate change and to assess its potential to flood low-lying cities and disrupt the agriculture and timber industries. The impact on the insurance business could be enormous, affecting the premiums policy holders pay, and altering decisions about where to locate new subdivisions, where to plant tree farms, and so on. This would begin to make climate change into an immediate economic issue.[28]

How rapidly this happens will be determined in part by governments, which since the late eighties have begun to take a serious interest in the climate problem. The most graphic symbol of this shift was the historic Framework Convention on Climate Change agreed to in Rio de Janeiro in 1992. The treaty notes that the long-run goal is to stabilize greenhouse gas concentrations in the atmosphere "at a level that would prevent dangerous anthropogenic interference with the climate system"— implying an eventual cut in emissions of at least 60–80 percent from the current level. With the convention entering into force in March 1994, the 159 signatories are required to submit a list of national sources of emissions by September 1994, while industrial countries also must develop national climate plans describing the actions they are taking to limit emissions.[29]

During negotiations, European governments had strongly argued for a binding freeze in industrial-country emissions, but determined opposition by President George Bush led to ambiguous language urging but not requiring such a target. The treaty, however, does include provision for adding to and strengthening it in later years, which in the Montreal ozone accords and other treaties has led to far faster progress than originally envisioned.[30]

Of the few national climate plans developed so far, most are limited to modest programs that increase funding of energy-saving projects, step up R&D on renewable energy technologies, and promote the use of natural gas. Few include the more politically difficult steps of reducing large subsidies that now encourage fossil fuel burning or assessing new energy or carbon taxes to discourage the use of those fuels. As a result, even with new plans in place, the United States, Japan, and the European Union—which together account for roughly 40 percent of the world total—all project small increases in their carbon emissions during the nineties. In the small countries of Denmark and the Netherlands, the climate issue has already had a more fundamental impact, however, spurring programs to promote energy efficiency, cogeneration, natural gas, and renewable energy, and initial efforts to increase energy taxes.[31]

The world is still at an early stage of dealing with the climate issue; future policies are likely to be driven by a combination of new scientific developments and actual weather events—as in 1988. Already, anecdotal evidence has emerged about aberrant weather affecting disparate corners of the globe in the past decade, ranging from record floods on the Mississippi River and violent wind storms in Northern Europe to deadly droughts in China, India, and southern Africa. As greenhouse gases continue to build in the atmosphere, the possibility looms that a series of catastrophic storms will devastate a large city or that a multiyear drought will wipe out a major crop, endangering world food supplies. If so, governments might well respond by accelerating the move to a new kind of energy system.

★   ★   ★   ★

Environmental limits have begun to shape both public and private choices about energy investments. At the simplest level, stricter environmental laws make it more expensive to build and operate energy facilities. The new reformulated gasoline being introduced in California in 1996, for example, is estimated to cost at least 10 percent more than the "regular unleaded" it replaces. In New Mexico, some 45 percent of the cost of the mammoth San Juan coal-fired power plant stems from environmental compliance. Many new nuclear and hydro construction projects have been halted by environmental objections, and governments have failed to provide the promised facilities for long-term storage of radioactive wastes, raising the possibility that some nuclear plants may be forced into early retirement.[32]

Although it is not yet officially regulated as a pollutant, carbon dioxide has emerged as a risk factor with a potentially enormous price tag. Moreover, since the carbon emissions associated with various fossil fuels and combustion technologies often correlate with emissions of regulated pollutants such as nitrogen and sulfur oxides, there is strong incentive to craft an integrated

abatement strategy that seeks to reduce use of the dirtiest fuels.

Take, for example, a decision about whether to build
a new oil refinery or coal-fired power plant that is projected to cost $1 billion and last 40 years. Air laws al-
ready require that such a facility have extensive pollution control equipment, but long before such an
investment were amortized the climate problem could
have reached the point where the government requires
the plant to be substantially modified or closed. Some
electric utilities in the United States have begun to consider this issue of "regulatory risk," and there is early
evidence that it has begun to shape their investment decisions.[33]

A growing number of economists and government
planners have argued in recent years that governments
should go a step further and levy taxes on fossil fuels
based on their carbon content—the so-called carbon
tax. Indeed, some wished to see such a tax in the climate
treaty itself. So far, most such proposals have foundered
on the shoals of energy industry politics, with the most
vehement opposition coming from the coal and oil industries. Efforts by the Clinton administration to
achieve a broad-based but small energy tax in 1993 were
beaten back by industry lobbyists, making it unlikely
that the world's largest fossil fuel user will adopt significant energy or carbon taxes anytime soon. Still, the idea
is far from dead. Several north European countries have
already adopted modest taxes on carbon, sulfur, and
other pollutants, and others are considering doing so.[34]

There are many other ways in which environmental
costs can be and are being incorporated in energy decisions. In the United States, many state regulatory commissions have ordered electric power companies to con-

sider "external" environmental costs in determining which option to pursue in providing electricity services. Massachusetts, for instance, has assessed environmental costs of $7,934 per ton of nitrogen oxides, $1,873 per ton of sulfur dioxide, and $26.45 per ton of carbon dioxide emitted by power plants. Including these costs would increase the delivered price of coal at a typical East Coast power plant by a factor of four. By mid-1993, six states were including such costs directly in their economic calculations, with many other states incorporating environmental costs more generally.[35]

Environmental costs have also been "internalized" by some of the air pollution laws enacted in the early nineties. The U.S. Clean Air Act Amendments of 1990, for example, set up a system of emission permit trading among the electric utilities and industrial plants required to reduce their sulfur dioxide emissions. Utilities that come in below the allowed emissions level are able to sell a permit to emit extra sulfur to companies that are over the limit. This puts a direct price on pollution; already, the permits can be bought and sold on the Chicago commodities exchange.[36]

In 1993, Southern California adopted a similar approach in order to meet the area's pollution targets—a Regional Clean Air Incentives Market. Under this program, 390 companies that are required to reduce their overall nitrogen oxide emissions by 75 percent and sulfur dioxide by 60 percent by 2003 will be permitted to buy and sell emission permits among themselves.[37]

As this example indicates, California remains at the forefront of governments forcing industry and consumers to take the environmental consequences of their energy decisions seriously. In 1989, the South Coast Air Quality Management District adopted a comprehensive plan with wide-ranging effects on life in the area. It includes the emissions trading program just described as well as a range of other programs to promote everything from carpooling to wider use of solar electric systems and fuel cells. At about the same time, the California Air Resources Board issued a requirement that 10 percent of the cars sold by major manufacturers in the state in 2003 be "zero emission vehicles," sparking a sudden acceleration in electric vehicle development around the world. (See Chapter 10.)[38]

Although the zero emission vehicle standard has been opposed as irrational and draconian by automakers, it has attracted strong political support from potential suppliers of new auto parts, and is clearly having an impact. As such, it illustrates but one of the new ways to encourage energy innovations that are now laying the groundwork for rapid change. Parts II and III of *Power Surge* describe many of the innovative developments now under way in response to those changes, and make the case for a quantum leap in the years ahead.[39]

II

The New Power
Brokers

4

Doing More with Less

In 1976, *Foreign Affairs* published an article entitled "Energy Strategy: The Road Not Taken?" written by an unknown 29-year-old named Amory Lovins. In it, the author sharply challenged some of the most fundamental assumptions used by energy planners, including the sacred principle of endless growth. To Lovins, who was trained in physics and had virtually no academic background in energy policy, the principal weakness in the current energy system was obvious: it was highly inefficient. The prevailing assumption that energy use had to rise in lockstep with the economy was, he claimed, mere superstition.[1]

To Amory Lovins, the answer to the world's energy problems was to make each unit of energy go further by increasing the efficiency of energy use. He suggested that with a modest effort, U.S. energy use in the nineties could be 30 percent below the levels being projected by experts. "The conventional view of the energy problem—where to get more energy, of any kind, from any source, at any price—was leading us to places we wouldn't want to be when we got there," Lovins said in 1982. "We were merely seeking to find, dig up, and burn fossil fuels faster and faster, a sort of 'strength through exhaustion'." Not only would supplies become scarcer and more expensive, but the implications of ever-expanding energy use in terms of pollution and disruption of communities would prove untenable.[2]

Lovins' stinging critique of the conventional view made him *persona non grata* to most energy planners. *Foreign Affairs* and other publications were flooded with letters criticizing him, and the U.S. Congress logged more than 2,000 pages of testimony and related submissions, including news items and editorials by authorities such as General Electric vice president Bertram Wolfe, former Secretary of Labor Peter Brennan, and Nobel-prize-winning physicist Hans Bethe. Wolfe described Lovins' vision as an "energy Shangri-la," while others called Lovins an "energy amateur" and a "purveyor of naked nonsense" that would lead to a "New Dark Age." At the same time, however, a growing band of environmentalists and consumer groups rallied to Lovins' cause, calling on governments to step up their investments in energy-efficient technologies.[3]

By the mid-nineties, even Lovins' severest critics had to admit he had become something of a prophet. Prompted by the rapid rise in oil prices during the seventies and eighties, businesses and individuals sought out cost-effective means to cut energy use—from adapting industrial processes and installing new equipment to adding insulation to home attics and tuning up basement boilers. Indeed, Lovins underestimated the speed of efficiency's gains: the Ford Foundation projected in 1973 that the country would use 140 exajoules of energy in 1993, and Lovins said that the figure would be 100, but the actual total turned out to be 91. During those 20 years, energy use rose only 15 percent while the economy as a whole expanded 57 percent.[4]

New thinking often requires someone who is not steeped in the existing paradigms. Lovins studied physics at Harvard while reading widely in a variety of fields, but never earned even a bachelor's degree—although at 21 years of age he was elected the youngest don at Oxford's Merton College in 400 years. Frustrated by the stultified academic environment, Lovins resigned that position in 1971 to take up a fight against a copper mine in his beloved Snowdonia National Park in northern Wales. The intervention proved successful, and not just in saving the park from development. There, he met David Brower, founder of Friends of the Earth (FOE), who convinced Lovins to bring his expertise to FOE to work on energy and other issues.

Instead of looking at the supply end of the spaghetti chart—a schematic that shows inputs, conversion, and final uses of energy—Lovins concentrated on the consumption side. Intuition told him that people do not want barrels of oil or watts of electricity, but warm houses, hot showers, and cold drinks. The service that energy provides is important; the actual amount used is not. His training in physics led him to focus on the incredible waste: only 15 percent of the energy in gasoline ever reaches the wheels of an average car, while just 30 percent of the energy in coal reaches electricity users. In Lovins' view, rather than travelling to the world's remote corners for new oil reserves, we should first seek out the less-expensive "oil fields" in our leaky attics and inefficient cars.

Lovins understood efficiency as a physical concept—one of the most fundamental characteristics of an energy system—and he avoided the confusion with energy conservation that was symbolized by President Jimmy Carter dressed in a cardigan sweater and sitting before a White House fireplace in 1977 to declare the moral equivalent of war on energy. The heart of Carter's speech, depicted by lowering thermostats and turning off unused lights, was his call for stringent efforts to conserve energy. Critics soon began to associate energy conservation with "freezing in the dark." Energy efficiency, on the other hand, is a physical measure of the ability of any given device to turn raw energy into useful work (gauged in miles per gallon for U.S. automobiles, for example). While conservation does have a role to play in reducing energy use, improved efficiency—that is, doing more with the same or less energy—is fundamental to enhancing economic well-being.

At the macroeconomic level, the overall energy efficiency of an economy can be defined by the amount of economic goods and services produced for every unit of energy used—a measure known as energy productivity. National energy productivity levels are influenced by a variety of factors, including the overall structure of the economy (the steel industry uses more energy than telecommunications, for instance) as well as the efficiency of individual technologies. Similar in concept to the much more widely followed trend of labor productivity, energy productivity is increasingly accepted as an indicator of economic progress. The relatively low level of energy productivity in the United States versus in Germany and Japan, for example, is considered a weakness. (See Figure 4–1.) Throughout the industrial era, energy productivity has generally increased, though the speed of improvement accelerated following the first oil crisis—rising 40 percent during 19 years in the United States and 46 percent in Japan.[5]

Still, the potential for increasing energy productivity remains large, and some energy analysts point enviously to the tenfold increase in labor productivity over the last century. Although reductions have already been made in the amount of energy used in buildings, industry, and transport, further improvements are within reach. Lighting provides a good example of the potential.[6]

Over the decades, new lighting technologies provided light using progressively smaller amounts of energy, evolving from wood fires to candles, whale oil, and later

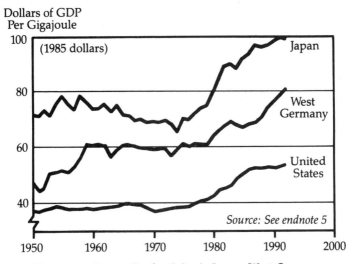

FIGURE 4-1. *Energy Productivity in Japan, West Germany, and United States, 1950–92*

kerosene lamps. The current offspring of Thomas Edison's incandescent bulb, invented in 1879, produce five times more lumens for each watt consumed than kerosene, while offering better-quality light. A second leap occurred in the early eighties when companies began to market a new type of energy-saving bulb: the compact fluorescent lamp (CFL)—a miniaturized version of the long fluorescent tubes that have been used for decades. (See Table 4–1.)[7]

The standard incandescent bulb uses more than 90 percent of the electrical energy it draws just keeping its tungsten filament hot enough to glow, meaning that it produces more heat than light. In contrast, a fluorescent bulb uses electricity to excite a tube-confined gas, which then radiates ultraviolet rays. Phosphors on the inner surface of the tube convert this radiation to visible light, producing much less heat in the process. As a result, a CFL is four times as efficient as an incandescent bulb, saving some $20 in electricity bills over its lifetime—even after including the additional cost of the lamp.[8]

As the lamps improved, CFL sales quadrupled between 1988 and 1993. Japan leads the way, with more than 80 percent of the country's home lighting now sup-

TABLE 4-1. *Efficacies of Selected Lighting Technologies*

Technology	Efficacy
	(lumens per watt)
Candle	0.2
Kerosene Mantle	0.8
Incandescent	14.7
Halogen	17.0
Compact Fluorescent	53.8

SOURCE: See endnote 7.

plied by compact fluorescents. Globally, 200 million CFLs were sold in 1993, compared with the roughly 9 billion incandescents purchased that year. CFLs last 10 times as long, however, so that they now effectively account for a healthy 15-percent share of the market for small bulbs.[9]

Similar gains have been or can be achieved in a host of energy-consuming areas, including buildings, automobiles, and industrial processes. Household appliances such as furnaces, water heaters, cooking ranges, and air conditioners all have improved markedly in recent years, yet the additional savings potential remains large. (See Table 4–2.) For U.S. refrigerators, electricity use was roughly cut in half between 1972 and 1992 as better insulation, more-efficient electric motors, and other relatively modest improvements were made. Additional changes could lower refrigerator energy use another 30–50 percent by the end of this decade. (See Figure 4–2.)[10]

TABLE 4-2. *Energy Efficiency Potential for U.S. Appliances*

| Technology | 1991/92 | | | |
	Stock	New	Best	Advanced
	(1985/86 models = 100)			
Refrigerator	125	165	210	300–770
Freezer	130	175	245	350–530
Central Air Conditioner	120	130	225	260–300
Electric Water Heater	105	120	335	400–500
Gas Furnace	110	130	150	150
Gas Water Heater	110	115	135	170
Gas Range	115	175	235	350

SOURCE: See endnote 10.

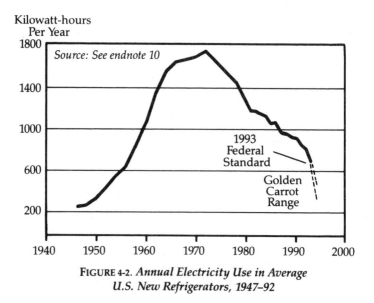

FIGURE 4-2. *Annual Electricity Use in Average
U.S. New Refrigerators, 1947–92*

As more advanced technology is developed and commercialized, the savings potential can keep growing. A
study by the American Council for an Energy-Efficient
Economy found that technologies likely to enter the
market during the next decade could further cut expected energy use in buildings by more than 25 percent.
In 1993, for example, computer manufacturers slashed
the energy use of desktop computers by roughly 70 percent simply by putting low-power microchips designed
for battery-powered notebook computers into their
desktop models. And a new kind of semiconductor chip
was unveiled that has the potential to cut dramatically
the power requirements of electric motors and increase
the carrying capacity of power lines.[11]

Although the energy use of individual appliances and
equipment can be cost-effectively slashed, the full im

pact of these innovations will be even greater once they are fully integrated into a new generation of buildings and motor vehicles designed to take advantage of the improved devices and materials. As described in Chapters 10 and 11, experimental automobiles and buildings designed in the early nineties suggest an ability to increase efficiency by factors of three to ten. Already, the fuel economy of new U.S. cars doubled between 1974 and 1985. Most of the improvement was achieved through changes in engines and design and through reductions in vehicle weight. Engine power output per volume increased by 36 percent between 1978 and 1987, for example. Yet a large potential remains: new technologies now under development suggest that cars run on internal combustion engines could be at least twice as efficient, and that more advanced technologies could permit yet another doubling in efficiency.[12]

★   ★   ★   ★

Industry claims a larger share of final world energy use than any other sector—roughly 45 percent. As energy costs rose in the seventies and eighties, many companies invested heavily in improved efficiency. In both the United States and Japan, real efficiency improvements raised the energy productivity of manufacturing by more than 37 percent between 1973 and 1988, saving companies billions of dollars. Yet a range of profitable opportunities to invest in improved efficiency still exist.[13]

Steelmaking illustates both the gains made and the potential to be tapped. Traditional open-hearth steel mills use vast amounts of coal to render iron ore and coke into steel, which later is fabricated into specific products, such as rolled plates and reinforcing bars. The multistep process creates large amounts of waste heat.

In integrated steel mills using continuous casting, first used on a large scale in Europe and Japan in the sixties, the steel does not need to be reheated for processing, reducing the losses. Large additional gains have been made since the seventies by shifting to electric arc steel minimills, which cut energy use by nearly half. Between 1975 and 1988, the U.S. steel industry reduced the amount of energy used to produce a ton of steel by 34 percent.[14]

Industry and government researchers believe that energy use in steelmaking could be cut an additional 42 percent within 25 years. Much of the savings would come from a continued transition to electric arc minimills and improvements in their operation. Through use of techniques such as oxygen-assisted melting of iron, or preheating of scrap using waste gases, power use in minimills can be cut by roughly 30 percent. An even newer technology, known as plasmamelt, is being tested in Sweden, and is expected to use just half as much energy as a minimill in producing steel from scrap.[15]

Nearly all industries depend on electric motors to provide drivepower in their processing lines, and the amount of electricity these use can be slashed by as much as 60 percent by using more energy-efficient motors and variable speed drives, and by better matching the size of the motor to the other drivetrain components. Even existing motors can be improved 10–15 percent through enhanced maintenance, including proper lubrication and rewinding.[16]

Many companies find that improving energy efficiency can have additional benefits. An ice cream factory in Framingham, Massachusetts, reduced its power use by a third, primarily by replacing a chlorofluorocarbon-based refrigeration system with an ammonia-based one, simultaneously lowering the use of ozone-depleting chemicals. Also, highly efficient electric motors were installed to run the homogenizing and pasteurizing processes. Overall, the electricity cost of producing a gallon of ice cream fell from 7.5¢ to 5.5¢. Another example can be found in the chlor-alkali industry, which produces two of the chemical industry's most common building blocks, chlorine and caustic soda. By using selective membrane cells instead of mercury ones to separate salt brine, power use is cut by 15–30 percent and toxic mercury is eliminated—which is why many companies have adopted this approach.[17]

One of the greatest opportunities for improving energy efficiency lies in cogeneration—the combined production of heat and electricity. In most such systems, the waste heat of power generation is made available for industrial processes or for heating a building. This allows up to 90 percent of the energy in a fuel to be used, far above the 33-percent figure found in many power plants. Efficient diesel and gas turbine generators that have appeared in recent years make even small cogenerators attractive. Responding to such advantages, as well as to reforms of utility laws, U.S. industry increased its cogeneration capacity from 10.5 gigawatts in 1979 to 37.1 gigawatts in 1991.[18]

Likewise, transmission and distribution of energy is undergoing rapid improvement, in part due to advances in solid-state electronics. Amorphous metal transformers, for example, use 70–90 percent less electricity than conventional iron core ones, a savings that is magnified by the fact that virtually all electric power passes through at least two transformers between the generating plant and the consumer. The compressor stations used to pump natural gas are also becoming steadily more efficient.[19]

The move toward more-efficient industrial processes

has been matched by a shift away from energy-intensive materials. A typical office building in the United States, for instance, uses one third as much steel as 30 years ago, and even greater reductions have been made in automobiles. In telecommunications, 1,000 kilograms of copper wire can now be replaced by just 25 kilograms of fiber-optic cable. Because of the reduced need for copper and the fact that fiber-optic lines have no resistance losses, the total amount of energy used in transmitting information can be lowered by 95 percent.[20]

★   ★   ★   ★

The importance of energy productivity is hard to overstate. Indeed, it is not a coincidence that countries with the highest levels, such as Japan and West Germany, enjoyed great economic success in the decade following the second oil crisis. Japan uses roughly 30 percent less energy in manufacturing a unit of output than the United States does (even after taking into account the differences in products), according to Lee Schipper and Stephen Meyers of the Lawrence Berkeley Laboratory in California. Countries with the least energy-efficient, most energy-intensive economies, such as Russia and Poland, were virtually bankrupted as soon as they were forced to pay the world market price for energy.[21]

Energy productivity is particularly important to developing countries. Because they are still in the early stages of building an industrial infrastructure, they have far greater latitude in using more-efficient processes and products—potentially following the path of Japan, which leapfrogged past the United States in the sixties by installing more-efficient industrial equipment. Unfortunately, many developing countries have so far failed to seize this advantage, sometimes even installing antiquated hand-me-down factories given to them by richer nations.

On average, developing nations saw their energy use expand 28 percent faster than their economies between 1973 and 1991, a trend that will need to be slowed or reversed if their development is not to be impeded by soaring energy bills. The potential is huge. A $7.5-million CFL factory in India, for example, could produce enough lamps to eliminate the need for $5.6 billion worth of coal-fired power plants. Each dollar invested in efficiency would save $740 in capital—before the fuel savings even begin. Replacing just one out of four incandescent bulbs in India would save some $430 million a year. Moreover, such a factory would also produce a valuable export product for which world demand is soaring.[22]

China has demonstrated that attention to improved energy efficiency makes sense even in poor nations. Starting in 1980, the government launched an ambitious program to improve energy use in major industries, directing about 10 percent of its total energy investment to efficiency. The nation cut its annual growth in overall energy use from 7 percent to 4 percent by 1985, without slowing growth in industrial production. Efficiency improvements accounted for more than 70 percent of the savings during 1980–85, with shifts toward less energy-intensive industries yielding most of the remainder. Had China failed to make such progress, energy use in 1992 would have been 82 percent higher than it actually was. Or—more likely—economic output would have grown far more slowly, as the country would have been unable to import the $53 billion worth of energy it needed. (See Figure 4–3.)[23]

For the world as a whole, the potential for energy sav-

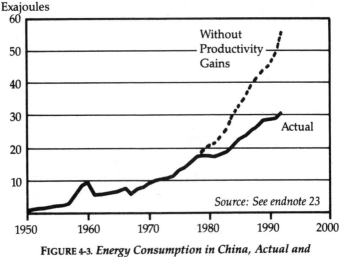

FIGURE 4-3. *Energy Consumption in China, Actual and Projected Without Productivity Gains, 1950–92*

ings over the next several decades is huge. Industrial countries such as the United States, Germany, and Japan could continue to expand their economies while reducing energy use by roughly 50 percent over the next 35 years. If the former Eastern bloc countries exploit their vast potential to improve efficiency, they will probably never use as much energy as they did in the eighties. And while developing countries will continue to use more energy, they could slice projected future growth in energy demand in half by investing in efficiency. Spending $10 billion annually on efficiency improvements worldwide would lead to gross average savings of $53 billion a year through 2025. Overall, global primary energy use by then could be kept just 42 percent above current levels.[24]

★   ★   ★   ★

Since 1976, energy efficiency has won many converts. Even the energy establishment, such as the World Energy Council and British Petroleum's former managing director Robert Malpas, now state that improving energy efficiency should be a centerpiece of future energy strategies. Yet their support for increased energy productivity often goes little beyond lip-service. Most governments have failed to eliminate many of the barriers to improving energy efficiency, and continue to underestimate the potential for future advances.[25]

The barriers to improved energy efficiency include entrenched special interests such as equipment manufacturers with a bias toward large supply-side projects, consumers who are ignorant of or unable to finance efficient alternatives, and engineers, architects, designers, and merchants who are currently given incentives to perform their jobs according to archaic specifications that downplay or ignore the importance of saving energy.

Engineering design fees, for example, are usually based on a percentage of the capital costs of a new building, which encourages engineers to install costly, oversized air conditioning and ventilation systems that lead to higher fees for the designer, and thus higher energy use. Such practices were often established in an age when oil was plentiful and cheap, and environmental problems overlooked. Also, consumers who generally pay for energy-conserving improvements in their homes face a far steeper interest rate than a power company that is borrowing money to build a new generating plant.[26]

Creative programs that have been pioneered in the last decade show a large potential for overcoming many of these barriers. The success in reducing the energy used in refrigerators offers an example of how governments can encourage change. California started the process in 1978 when it issued standards decreeing maximum levels of electricity use for refrigerators, requiring reductions of 20 percent for the average model. The effect of California's law was felt nationwide, as manufacturers realized it was not economical to produce special refrigerators for a state that by itself accounted for one fifth of the nation's sales. After companies quickly achieved the required reduction, California tightened the standard. And when other states followed with standards of their own, manufacturers begged a resistant Reagan administration for a single national standard. By 1987, national standards for refrigerators and other major appliances became law.[27]

Although standards set a floor on the efficiency of new devices, they give companies little incentive to make further improvements. Faced with this dilemma, David Goldstein, a physicist with the Natural Resources Defense Council, came up with the idea of using carrots as well as sticks to encourage manufacturers. He reasoned that producers could be convinced to build better refrigerators if they were given a financial incentive to do so.[28]

His idea was put to the test in Sweden, where in 1989 Hans Nilsson, director of the government's Department of Energy Efficiency, launched a European competition for an efficient refrigerator. Key to stoking interest was a government guarantee to manufacturers that the winner would receive large purchasing contracts from housing companies that agreed to be part of the program. The winning unit, made by the Swedish company Electrolux, slashed energy use by an additional 35 percent, exceeding organizers' goals. Electrolux started marketing its new model in late 1991, and though sales were initially slower than hoped for in Sweden, they accounted for 60 percent of the company's sales in Germany and 80 percent in Finland by early 1993. Moreover, other manufacturers have started marketing more-efficient refrigerators to capture some of Electrolux's newfound market. Emboldened by their success, Nilsson and his colleagues have since completed competitions for advanced windows, lighting, and other technologies.[29]

Meanwhile, a more ambitious effort was gearing up in the United States. Twenty-four electric utilities, with support from the U.S. Environmental Protection Agency (EPA), state utility regulators, and environmentalists including Goldstein, pooled their resources to launch their own refrigerator competition—known as the Golden Carrot. In a winner-take-all contest, the manufacturer who marketed the most cost-effective refrigerator that exceeded federal standards by at least 25 percent while eliminating the use of ozone-depleting chlorofluorocarbons was guaranteed a cash prize of $30 million. Whirlpool Corporation won the competition with an appliance that uses 30 percent less electricity than required under 1993 federal standards. A second round of competitions is planned for refrigerators as well as other appliances—including clothes washers and commercial-scale air conditioners.[30]

The Golden Carrot competition is only one of many incentive programs, including some aimed at improving energy use in buildings, computers, and lighting, that EPA has launched. The programs, along with those in Sweden, show the enormous potential to jumpstart the market for energy-efficient appliances. And efficient technologies can move into the market far more rapidly than energy-producing ones: In 1974, construction started on the second unit of the Comanche Peak nuclear power plant, a reactor that was finally ready to generate power in 1993. In contrast, the first CFL was not sold until 1982, yet the CFLs sold a decade later displaced the equivalent of five nuclear power plants the size of Comanche Peak's 1,150 megawatts.[31]

Today, Amory Lovins is the research director of the Rocky Mountain Institute in Snowmass, Colorado, and still stumps widely for improved energy productivity. But now he carries the latest compact fluorescent lamps—and other demo models—in his briefcase. And his audience and message have broadened. In fact, Lovins is now paid considerable sums to advise his former critics, including utility executives and government officials, on how to overcome the obstacles to improving energy efficiency.[32]

Lovins has remained faithful to his original thesis, and gradually the rest of the world is catching up. Higher energy productivity is increasingly viewed as key to reducing the environmental burdens of today's fuels, as well as to improving the economic competitiveness of many industries. As later chapters will argue, however, efficiency's importance goes even further: by providing a highly efficient energy infrastructure—from power plants to light bulbs—these new technologies will one day make it possible to run our factories on solar energy, and to fill our gas tanks with hydrogen.

5

Prince of the Hydrocarbons

When a U.S. Senate hearing was called in 1984 to assess the prospects for natural gas, almost everyone expected a gloomy session. At the time, gas production in the United States had been falling for 12 years, and prices had tripled in a decade. This seemed a textbook example of a rapidly depleting resource—a daunting prospect given that at its peak natural gas had met more than one quarter of U.S. energy needs. In Europe, the situation was similar: gas reserves appeared to be dwindling, and governments put legal restrictions on its use. Few were surprised, then, when Charles B. Wheeler, Senior Vice President at Exxon—the world's largest oil company— told the Senate that natural gas was essentially finished as a major energy source: "We project a shortfall of economically available gas from any source," he declared.[1]

Only one voice interrupted the gloom that pervaded the hearing room: that of Robert Hefner, an iconoclastic geologist who headed a small Oklahoma gas exploration company and who was grandson of one of the earliest oil wildcatters. Hefner's words echoed in the hearing room: "My lifetime of work requires that I respectfully have to disagree with everything Exxon says on the natural gas resource base."[2]

A decade later, legions of government and industry analysts have had to eat their words, while Hefner has turned his contrarian views on natural gas into a comfortable fortune. U.S. gas prices fell sharply after 1986, and by 1993, production had climbed by 15 percent. At the global level, world gas production has risen 30 percent since the mid-eighties, with increases recorded in nearly every major country.[3]

In fact, growing signs suggest that the world is already in the early stages of a natural gas boom that could profoundly shape our energy future. If natural gas production can be doubled or tripled in the next few decades, as Hefner and a growing number of geologists believe, this relatively clean and versatile hydrocarbon could replace large amounts of oil and coal. Because it is easy to transport and use—even in small, decentralized technologies—natural gas could help accelerate the trend toward the more efficient energy system described in Chapter 4 and, over the long run, the transition to renewable sources of energy.

★   ★   ★   ★

The Chinese were the first to use geologically occurring methane gas, approximately 2,000 years ago. Gas was then ignored until the early twentieth century, when it reappeared as a by-product of oil extraction that was often vented. Many major cities had developed facilities to gasify coal (and occasionally oil), which was then distributed to homes for lighting and other purposes. "Natural" gas, it was soon discovered, was less expensive and cleaner burning than coal gas, and it captured the market in one region after another—pushing coal gas out of Western Europe, for example, in the sixties, after the discovery of extensive gas fields in the Netherlands. By the seventies, natural gas was a major fuel for homes and industry and was used as feedstock in the production of a wide range of petrochemicals.[4]

Like electricity, natural gas required an extensive transmission and distribution system before it could become a part of the energy system, ranging from facilities to gather and separate gas to compressor stations, long-distance pipelines, and millions of kilometers of distribution pipes reaching out to individual homes. Much of the United States, from California to the Midwest, was hooked into an interstate network of pipelines during the fifties and sixties, a network that finally reached corners of the northeast and Florida in the eighties.[5]

In Europe, the gas resources of the Netherlands were supplemented in the seventies by long-distance pipelines from as far away as Algeria and Siberia, and in the eighties by a multibillion-dollar network of undersea pipelines extending thousands of kilometers into the North Sea. In Japan, which has little indigenous oil or gas, large-scale use began in the seventies with the development of a system for importing liquefied natural gas (LNG) by ship from Indonesia and Malaysia.[6]

The environmental advantages of natural gas over other fossil fuels were a strong selling point from the start. Methane is the simplest of hydrocarbons—a carbon atom surrounded by four hydrogen atoms—with a higher ratio of hydrogen to carbon than other fossil fuels. Natural gas helped reduce the dangerous levels of sulfur and particulates in London's air in the fifties. In fact, these two contaminants are largely absent from natural gas by the time it goes through a separation plant and reaches customers. Natural gas combustion also produces no ash and smaller quantities of volatile hydrocarbons, carbon monoxide, and nitrogen oxides than oil or coal do. And, unlike coal, gas has no heavy metals.[7]

As a gaseous fuel, methane tends to be combusted more thoroughly than solids or liquids are. Due to its lower carbon content, natural gas produces 30 percent less carbon dioxide per unit of energy than oil does, and 43 percent less than coal, reducing its impact on the atmosphere. It is also relatively easy to process compared with oil, and less expensive to transport—via pipeline—than coal, which generally moves by rail. Methane gas is not entirely benign, however. When not properly handled, it can explode. And as a powerful greenhouse gas in its own right, it can contribute to the warming of the atmosphere. But with careful handling, both these problems can be reduced dramatically.[8]

From 1950 to the mid-seventies, natural gas was the world's fastest growing major fuel, with use rising sevenfold during the period. Growth slowed dramatically in the next decade, however, led by a decline in the use of gas in the United States, which had been the world leader. Following several years of rising prices and falling sales, Harvard Business School's *Energy Future*, published in 1979, concluded that "the nation should not plan on greater quantities of natural gas to stop the rise in oil imports." At about the same time, the U.S. Congress passed a law forbidding the use of natural gas in new power plants, forcing utilities to switch to coal.[9]

The demise of natural gas during this period turned out to be rooted in a bizarre system of price regulations that slowly strangled the industry. Starting in the fifties, the price of gas traded across state lines was strictly controlled by the federal government, which helped spur demand but which reduced the incentive for producers. At first, there was plenty of gas available, simply as a by-product of oil exploration, and it was extremely cheap to produce. But as oil production peaked, and supplies of this "associated" gas dwindled in the seventies, gas reserves began to drop.[10]

The U.S. Congress and the Carter administration responded by lifting natural gas price controls, but it took much of the eighties for them to disappear. Meanwhile, the transition created problems of its own: as controls were lifted on newly discovered reserves, prices skyrocketed, causing a decline in demand and deepening the perception of scarcity. Next, falling demand created a "bubble" of excess supplies, pushing the average price down by more than 40 percent between 1984 and 1990—to an oil equivalent of just $9 per barrel.[11]

Although lower prices reduced the incentive to find gas, by the late eighties the price of most of the gas sold had been decontrolled and the industry had become vigorously competitive, with open access to the large interstate pipelines and an active spot and futures market. As a result, U.S. gas use rose 24 percent between 1986 and 1993, and prices strengthened to the equivalent of about $12 per barrel. Although Canadian producers were able to capture more than a third of the increased demand, U.S gas resources turned out to be much more abundant than expected. Production had increased 15 percent by 1993, while that of oil—which was selling at a substantially higher price—continued to fall. (See Figure 5–1.)[12]

The late eighties and early nineties were marked by expanding production and use of natural gas on every continent except Antarctica, including in many countries that had hardly used it in the past. Argentina, Egypt, Kazakhstan, Malaysia, Mexico, Turkey, and Turkmenistan are among the developing countries working to increase their reliance on natural gas. New pipelines may soon allow Bolivia to sell gas to Brazil, and Myanmar to sell it to Thailand. Even in the nineties, however, many nations, including several in the Middle East and Central Asia, are still wasting large amounts of gas. In Nigeria, for example, the gas that was flared off by the oil industry in 1990 could have met all the nation's energy needs as well as those of five of its neighbors.[13]

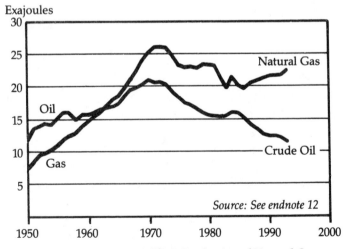

FIGURE 5-1. *Continental U.S. Production of Natural Gas and Crude Oil, 1950–93*

In Europe, major new pipelines are being extended from Norway, Russia, and northern Africa, and use is rising particularly rapidly in eastern Europe, where gas is replacing coal in household furnaces and district heating systems. China, which relies minimally on natural gas today, is also stepping up its exploration efforts, and building major new pipelines in Guangdong and Sichuan. Elsewhere in Asia, trade in liquefied natural gas is expanding rapidly, and may soon reach India and Pakistan.[14]

Qatar, whose Middle Eastern oil fields are starting to run dry, is even considering selling LNG to Israel. There is also talk of extending an undersea pipeline from the rich gas reserves of the Middle East to the 1.2 billion consumers on the Indian subcontinent. Japan, meanwhile, is considering a domestic pipeline system, including undersea connections to the Russian oil fields of Sakhalin Island, and possibly to southeast Asia. This project might well open the way for a Far Eastern pipeline network running from Siberia to Australia.[15]

Natural gas is far more versatile than either coal or oil, and with a little effort can be used in more than 90 percent of energy applications. Yet until recently its use has been largely restricted to household and industrial markets in which it has thrived. In North America, for example, natural gas is far and away the most popular heating fuel; by the early nineties, nearly two thirds of the single family homes and apartment buildings built in the United States had such heating systems. In recent years, new technologies such as gas-powered cooling systems and heat pumps have even allowed this energy source to challenge electricity's dominance of additional residential and commercial applications. More significantly, natural gas has begun to find its way into energy markets from which it was excluded in the past, including transportation and electricity generation.[16]

Gas has always been an attractive fuel for electric power generation, but high prices and legal strictures deterred its use by utilities during the seventies and eighties, so most of the plants built then were fueled by coal or nuclear power. (By 1990, gas constituted only 8 percent of the fuel used in electricity generation in North America, and 7 percent in Europe.) Until recently, most power plants, whatever the fuel, used a simple Rankine cycle steam turbine: the heat generated by burning the fuel produced steam, which spun a turbine connected to an electricity generator. Although this technology had progressed steadily for decades, by the sixties its efficiency in turning the chemical energy of fossil fuels into electricity had levelled off at about 33 percent, meaning that nearly two thirds of the energy was still dissipated as waste heat. The inefficiency of this process made it desirable to use as cheap a fuel as possible, and in an era when environmental constraints were minimal, coal was the obvious choice.[17]

Although natural gas can be and is used in simple steam-cycle power plants, its gaseous nature makes it attractive for fueling a more recent technology: the jet engine. First developed during World War II and then commercialized in the fifties as a result of military R&D programs, jet engines consist of rotating blades into which pressurized fuel is injected and then combusted, spinning the turbine and propelling an aircraft forward. When mounted on the ground and attached to a generator, the device is known as a gas turbine. During the sixties and seventies, natural-gas-fired turbines began to serve an important role in the power industry, producing electricity during periods of peak power demand. At the time, however, the generators were less than 30 percent efficient in producing electricity, limiting their economic appeal.[18]

This situation changed dramatically in the late eighties as natural gas prices fell and turbine technologies improved. Much of the recent gas turbine renaissance is focused on the combined-cycle plant—an arrangement in which the excess heat from a gas turbine is used to power a steam turbine, boosting efficiency. Combined-cycle plants reached efficiencies of more than 40 percent in the late eighties, with the figure climbing to 50 percent for a General Electric (GE) plant opened in South Korea in 1993. At about the same time, Asea Brown Boveri announced plans for a 53-percent-efficient combined-cycle plant. These generators are inexpensive to build (roughly $700 per kilowatt, or a little more than half as much as the average coal plant) and can be constructed rapidly—two-and-a-half years for the huge 1,875-megawatt Teeside station completed in the United Kingdom in 1992.[19]

More versatile turbines are also on the way. Using the advanced metals, new blade designs, and high compression ratios of jet engines, engineers are producing smaller "aeroderivative" turbines. By applying a host of modest innovations such as steam injection, the efficiency of these one-cycle devices reached 39 percent and is expected to close in on 60 percent by 2000. An even more advanced turbine called a "ram jet" may one day achieve electrical conversion efficiencies of more than 70 percent, according to some engineers. Because aero turbines are built in factories, costs are as low as $350 per kilowatt, and they can be installed in just a few months. Moreover, they range in size from 52,000 kilowatts for an adapted DC-10 aircraft engine to less than 20 kilowatts for turbines being developed for possible use in automobiles. Natural-gas-powered turbines and engines are helping to drive the growing use of "combined heat and power" systems, in which the waste heat from power generation is used in factories, district heating systems, or even individual buildings. Small-scale cogeneration has already become popular in Denmark and other parts of northern Europe.[20]

Gas turbine plants have major environmental advantages over conventional oil or coal plants, including no emissions of sulfur and negligible emissions of particulates. Nitrogen oxide emissions can be cut by 90 percent and carbon dioxide by 60 percent. (See Table 5–1.) Indeed, the combination of low cost and low emissions has allowed natural gas to dominate the market for new power plants in the United States and the United Kingdom during the early nineties; even larger markets are unfolding in South Asia, the Far East, and Latin America. In the future, this technology could spur utilities to convert hundreds of aging coal plants into gas-burning combined-cycle plants—for as little as $300 per kilowatt. Worldwide, some 400,000 megawatts' worth of gas turbine plants could be built by 2005, according to a GE forecast; units are already up and running in countries as diverse as Austria, Egypt, Japan, and Nigeria. A secondary result of this boom is the emergence of natural gas as the dominant fuel for new power plants in many countries.[21]

In the future, a device called a fuel cell offers an even more versatile and efficient means of converting natural gas to electricity. First identified by scientists in the early nineteenth century, fuel cells are electrochemical devices consisting of an electrolyte and two electrodes that generate an electric current by combining hydrogen and oxygen ions—avoiding entirely the electromechanical

TABLE 5-1. *Conversion Efficiencies and Air Pollutants, Various Electricity-Generating Technologies[1]*

Technology[2]	Conversion Efficiency[3]	Emissions		
		NO_x	SO_2	CO_2
	(percent)	(grams per kilowatt-hour)		
Pulverized Coal-Fired Steam Plant (without scrubbers)	36	1.29	17.2	884
Pulverized Coal-Fired Steam Plant (with scrubbers)	36	1.29	0.86	884
Fluidized Bed Coal-Fired Steam Plant	37	0.42	0.84	861
Integrated Gasification Combined-Cycle Plant (coal gasification)	42	0.11	0.30	758
Phosphoric Acid Fuel Cell (using hydrogen reformed from natural gas)	36	0.04	0.00	509
Aeroderivative Gas Turbine	39	0.23	0.00	470
Combined-Cycle Gas Turbine	53	0.10	0.00	345

[1]Data are for particular plants that are representative of ones in operation or under development. [2]Coal plants are burning coal with 2.2 percent sulfur content. [3]Higher heating values, which give lower efficiency levels, are used throughout.

SOURCE: See endnote 21.

generators of today's power plants. The big boost for these came in the sixties, when space scientists began looking for a small, self-contained power system and wound up investing billions of dollars in fuel cells that were successfully used on spacecraft.[22]

The fuel cell has three major advantages over conventional power generators: It is relatively efficient, converting between 35 and 65 percent of the energy potential of a fuel such as methane or hydrogen into electricity. It produces less air pollution than a conventional generator due to its lack of combustion and greater efficiency. And it is virtually silent. Fuel cells can be as big as a large conventional power plant or small enough to fit into a space shuttle—or under the hood of a car. A U.S. company called International Fuel Cells is the world's largest manufacturer, and is planning to turn out 200 of its 200-kilowatt phosphoric acid fuel cells annually in the nineties. Several other types of fuel cells are under development, mainly in the United States and Japan.[23]

Already, Tokyo Electric Power and Southern California Gas have installed fuel cells that provide power and heat in hospitals, hotels, office buildings, and other commercial facilities. At roughly $3,000 per kilowatt in 1994, fuel cells are more expensive than conventional power plants, but their cost is projected to decline as they move into mass production in the late nineties and beyond. Two decades from now, new buildings could have natural-gas-powered fuel cells in their basements that would not only generate electricity but also replace today's furnace, water heater, and air conditioner.[24]

Natural gas is even beginning to break oil's stranglehold on the transportation market. Compared with gasoline and diesel fuel, natural gas has both economic and environmental advantages. In the United States, for instance, its wholesale price was less than half that of gasoline in 1993, a disparity caused in part by the cost of refining gasoline. As in other applications, the chemical simplicity of methane is a major advantage, reducing emissions and allowing for less engine maintenance. Until recently, compressed-gas vehicles were confined to just a few countries: nearly 300,000 are found on

Italy's roads, and more than 100,000 on New Zealand's.[25]

The main challenge in using natural gas in motor vehicles lies in storing the fuel in the car—usually in cylindrical pressurized tanks. The early tanks were bulky and heavy, but manufacturers are now producing lightweight cylinders made of composite materials that will make it possible to build virtually any kind of natural gas vehicle with a range similar to a gasoline-powered one. Engineers believe they can cleverly design a tank into the smallest passenger cars without sacrificing any trunk space.[26]

Switching to natural gas will be even easier for buses, trucks, and locomotives, as their size means that finding room for the tanks is not an issue. Many local bus systems are already switching over, in order to avoid the cancer-causing particulates and other pollutants that flow from the diesel-powered engines currently used. Operators of local delivery vehicles are moving in the same direction. The United Parcel Service in the United States, for example, is testing natural gas in its vehicles. The idea of switching train locomotives from the currently dominant diesel-electric systems to gas-electric ones is just beginning to be studied. In the United States, Union Pacific and Burlington Northern are both testing the use of LNG in their engines. Preliminary data indicate favorable economics and excellent environmental performance.[27]

Converting service stations so that they can provide natural gas is also straightforward, and several oil companies have begun to do this. In Europe and North America, virtually all cities and many rural areas have gas pipes running under almost every street, and simply need to provide service stations with compressors for putting the gas into pressurized tanks. And it may well be possible for residential buildings, millions of which are already hooked up to gas lines, to be fitted with compressors, meaning fewer trips to a service station. As of early 1994, about 900 U.S. service stations were selling natural gas, with four to five more joining their ranks each week.[28]

Vehicles running on natural gas are likely to have far lower emissions than gasoline-powered cars, though engines, fuel injection systems, and catalytic converters must be redesigned for the full environmental gains to be realized. Preliminary data suggest that with such adjustments, emissions of carbon monoxide, reactive hydrocarbons, and particulates would be greatly reduced. Chrysler began building a natural-gas-fueled mini-van in 1993 that was the first vehicle to meet California's ultra-low emission standard. Moreover, natural gas can be burned in a high-compression engine, which allows efficiency to improve 15–20 percent and carbon dioxide emissions to be cut 13–17 percent. And because their fuel systems are sealed, these vehicles produce none of the evaporative emissions that flow from the fuel tanks and lines of gasoline-fueled vehicles.[29]

As with gas turbines, the main pollution problem with hot-burning natural gas engines is nitrogen oxide emissions. Since burning natural gas yields a much smaller range of emissions than gasoline does, these engines will require a different though ultimately simpler and probably more effective kind of catalyst. Engineers believe that a special catalytic converter can nearly eliminate nitrogen oxide emissions and at the same time keep hydrocarbon emissions low. Natural gas automobiles do emit some methane—itself a greenhouse gas—but with care, emissions can be controlled.[30]

Interest in natural gas as a vehicle fuel blossomed in the early nineties as cities such as São Paulo and Mexico City struggled to cope with intractable air pollution. In the United States, many state and local governments began to promote natural gas vehicles in public and private fleets, while car manufacturers built gas-powered versions of some of their auto and light truck models, and gas distribution companies converted gasoline-powered cars to the use of natural gas. In many regions, natural gas has eclipsed both ethanol and methanol, the two new automotive fuels that commanded most of the attention in the eighties. An industry study estimates that as many as 4 million natural gas vehicles could be on U.S. roads by 2005.[31]

*   *   *   *

From its earliest days, petroleum geology has been an inexact science at best. The first oil fields were found by accident, and for decades thereafter most reserves were found by a "hunt-and-drill" method that left many dry holes before anything of significance was found. In the days when gushers were being struck at places like Teapot Dome, Titusville, and the shores of the Persian Gulf, this technique sufficed. And over time, enough basic geologic knowledge was acquired to provide at least some indication as to whether a particular formation of rock had the capacity to contain liquid or gaseous hydrocarbons. There is still much to be learned about oil resources, however, and far more to be learned about natural gas, which in many respects remains the poor stepchild of the oil industry.

Even in the mid-nineties, geologists disagree vehemently about how much natural gas remains to be found. Nevertheless, the trend is clear: as knowledge grows, the estimated size of the resource base expands with it. The U.S. experience provides insights, since it has the most extensive gas industry, and its resources are the most heavily exploited. The sharp increase in U.S. gas production since the mid-eighties has been accompanied by a reevaluation of the resource base. A 1991 National Research Council study of official estimates made by the U.S. Geological Survey (USGS) found that "after a detailed examination of [USGS's] data bases, geological methods, and statistical methods, the committee judged that there may have been a systematic bias toward overly conservative estimates." In some cases, estimators would not count a reservoir of gas until it was virtually out of the ground.[32]

This bias toward underestimation is even more pronounced in other parts of the world, where few non-oil-producing areas have been explored for gas and some have been examined only superficially. In many areas, wells that produce only gas are declared "unproductive" and then plugged. In contrast to oil, natural gas is often more difficult to find; liquid petroleum only exists at modest temperatures and pressures, limiting it mainly to shallow deposits, while gaseous methane does not readily break down and can be found at extreme temperatures and depths. Natural gas is now sometimes extracted from reservoirs four to six kilometers beneath the surface, where large quantities can be found in small spaces.

As with virtually all other energy technologies, the techniques for locating and developing new gas fields are advancing rapidly—propelled in part by the advent of computer software that makes it possible to generate three-dimensional seismic images of the subsurface geology, and to determine how much gas may be there. As a result of these and other developments, the real cost of finding and extracting gas has declined markedly since the mid-eighties, and a new category of "tight" gas has become economically accessible. Natural gas is also being found in abundance in many coal seams that were not thought to be exploitable.[33]

The recent trend in gas resource estimates is well illustrated by the U.S. experience. Even the more conservative estimates used by the USGS—showing 700 exajoules—are more than double those used by Exxon in 1984, and organizations that take a longer-range perspective have come up with much higher numbers. The National Petroleum Council, for example, estimated in 1992 that the United States has 1,400 exajoules of additional natural gas that can be economically exploited with technologies that will be available by 2000. (See Figure 5–2.) This is enough to last the country 60 years at current rates of use. It is estimated that most of the gas can be found and extracted at a price of $20 per barrel of oil equivalent ($3.30 per gigajoule) or less.[34]

U.S. gas resources are only a tiny fraction of the world total, and discoveries are now proceeding more rapidly in other regions. On most continents, more natural gas than oil already has been identified, a gap that will widen as exploration accelerates. During the past two decades, enormous amounts have been discovered in Argentina, Indonesia, Mexico, North Africa, and the North Sea, among other areas. Each region either is or could become a major exporter of natural gas. In addition, some of the former Soviet republics in central Asia have extensive gas resources, which are relatively inaccessible but are being studied by major western oil and gas companies.[35]

Russia is one of the keys to the global gas outlook. It is

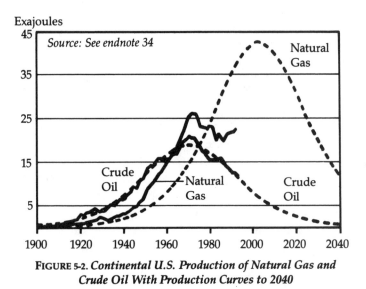

Exajoules

FIGURE 5-2. *Continental U.S. Production of Natural Gas and Crude Oil With Production Curves to 2040*

the largest producer and has the most identified reserves. While oil production has declined catastrophically with the collapse of the communist economic system, the flow of gas has fallen only slightly. Western experts have reassessed the Russian data and decided that the gas fields are even richer than previously believed. According to "mean" estimates by USGS scientists, the total Russian resource base is close to 5,000 exajoules—enough to meet current world demand for 60 years. Because it is located in Siberia and other remote areas, much of this gas must be moved a long distance. But it is still within reach of more than half the world's energy consumers, including the 1.8 billion who live in China, Japan, and Europe.[36]

At least 50 additional countries have natural gas reserves that are minor on a global scale but sufficient to fuel their economies for decades. (See Table 5–2.) Most developing nations have barely begun to look for methane, however. One reason is that the large oil companies with capital and technical know-how are less interested in gas since it is not easily sold for hard currency. Moreover, few developing countries have made the investments in pipelines and distribution systems needed for

TABLE 5-2. *Proven Natural Gas Reserves by Country, 1990*

Country	Natural Gas Reserves	Gas Reserves Remaining at Current Rates of Oil Use
	(exajoules)	(years)
Soviet Union	2,100	95
Iran	700	437
Argentina	300	282
Canada	300	100
United Arab Emirates	225	659
Qatar	200	2,080
Saudi Arabia	200	906
United States	200	7
Algeria	125	132
Iraq	125	117
Venezuela	125	56
Norway	100	207
Indonesia	100	64
Australia	90	75
Mexico	85	29
Malaysia	60	176
Kuwait	55	69
Libya	50	84
India	45	19
China	40	9

SOURCE: See endnote 37.

widespread use of gas, and the World Bank traditionally has been stingy in its support of such projects. In Mexico, the natural gas industry is still monopolized by the state oil company, Pemex, which in 1990 drilled just four natural gas wells (compared with 7,170 in the United States), despite the official policy to increase the country's reliance on natural gas. Once Mexico and other countries get serious about this energy source and open up the industry to competition, proven reserves and production are likely to grow rapidly.[37]

China, which according to official statistics has less than 1 percent of the world's proven gas reserves, may be an important example of hidden potential. The nation's energy planners, who in the fifties and sixties overtapped and badly damaged some of their shallow gas reserves, have been so convinced of its limited potential that they have based the country's development on coal. Yet China has recently stepped up exploration efforts in several areas, and is finding extensive resources.

Robert Hefner, who explored for gas in China in the eighties, believes that China—about the same geographical size as the United States—may have about the same amount of gas. (U.S. gas production today would be sufficient to meet all of China's current energy needs if it were used at U.S. efficiency levels.) Some of the largest deposits may be in the Ordos and Sichuan basins, within easy reach of most of the country's industrial and population centers. Although the costs of building a gas transmission system are considerable, they would be offset by reduced investments in coal mines, rail lines, and power plants, as well as by avoided ecological damage.[38]

Estimating the ultimate scale of the worldwide resource of natural gas is still difficult, but better information is becoming available all the time. One thing is clear: the "proven reserves" figures published each year by government and international agencies—now set at 5,400 exajoules—are but the tip of the iceberg. Unlike coal, gas is relatively expensive to find but cheap to produce, so companies have no incentive to identify more than they will need in the next few years. Broader estimates by the U.S. Geological Survey show 12,000 exajoules of "conventional" resources that are economically accessible using today's technologies. This represents 145 years of supply at the current level of world production.[39]

Many geologists, including Robert Hefner, believe that even these figures underestimate the resource base by a factor of three or more, since they exclude potential gas in many unexplored areas, as well as much of the "tight" gas, "deep" gas, and "coalbed" gas that may exist in many parts of the world. They point out that methane gas is relatively ubiquitous in many subsurface formations, including massive but currently hard to reach methane hydrates that are found on the ocean floors. And they argue that the oil bias of most petroleum geologists and the relatively primitive state of geological understanding have led to profound misperceptions about the size of the resource base.[40]

Geologist M. King Hubbert demonstrated in the sixties that if you know the size of a total resource base, you can project a logistic production curve—following a classic bell shape. Using this technique, he was able to predict the peaking of U.S. oil production with surprising accuracy. World natural gas production has followed a relatively steady growth curve so far, but the size of the logistic curve that it ultimately will follow depends on the scale of the resource. If the conservative USGS numbers are right, production is likely to roughly double the current level before peaking, but if the resource is 50 percent larger, production might triple before peaking— probably between 2030 and 2040. (See Figure 5–3.) Given the conservative nature of the USGS estimates and the rapid increase in global resource evaluations as a result of improved technology and more exploration, even the higher number seems fairly modest.[41]

Based on this analysis, and assuming that the economic and environmental advantages of this energy source continue to stimulate demand, a major boom in the use of natural gas can be expected during the next few decades. Since world oil production is likely to grow only modestly, and could well begin to decline soon after the turn of the century, natural gas will almost certainly soon be the most important fossil fuel—available in sufficient quantity to replace many existing uses of oil and coal.[42]

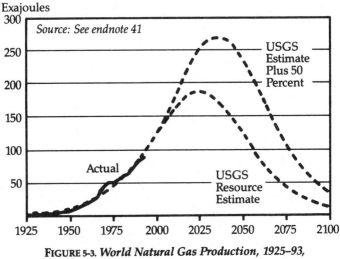

FIGURE 5-3. *World Natural Gas Production, 1925–93, With Production Curves to 2100*

Environmental factors—not resource limits—may ultimately constrain world reliance on this hydrocarbon. So far, natural gas has accounted for just 10 percent of the total carbon dioxide vented to the atmosphere since 1860, but eventually, as it supplants oil and coal, it could become the largest producer of this greenhouse gas. If the resources prove as abundant as some geologists now believe, gas may eventually follow the fate of coal—phased out for environmental reasons long before the resource is exhausted.[43]

One concern with methane as an energy source is that it is a powerful heat-trapping gas in its own right—building rapidly in the atmosphere in recent decades, and contributing to climate change. If rising use of gas were to increase leakages of methane, global warming would be exacerbated and the benefits of reduced carbon emissions partly offset. But natural gas is only one of many sources of methane buildup, along with coal and oil extraction, landfills, rice cultivation, and livestock. Scientists estimate that natural gas accounts for 25–42 percent of the methane emissions that stem from fossil fuels, or up to 12 percent of total emissions related to human activities.[44]

The key to sustainable reliance on natural gas is to reduce the amount of leaked methane at the same time that use of the fuel is increased. Anecdotal evidence suggests that this can be achieved relatively easily. For example, the U.S. Environmental Protection Agency and the gas industry have come up with a plan for substantially reducing losses in the relatively tight U.S. system. The priority, however, is to repair the notoriously porous gas systems of the former Soviet Union, where leakage rates are estimated at up to 10 times those in the West. Since 1992, the gradual decontrol of Russian gas prices has spurred efforts to improve the efficiency of the distribution system; western companies and governments have begun to contribute to these efforts—in some cases being paid in saved gas rather than the less-stable Russian ruble.[45]

These efforts may already be having an impact on atmospheric concentrations of methane, which were reported to have nearly stabilized between mid-1992 and late 1993. Methane lasts an average of just 10 years in the atmosphere; further reductions in emissions could lower atmospheric concentrations substantially over the next two decades—something that would take nearly a century to accomplish for carbon dioxide.[46]

Even as reliance on natural gas grows during the next few decades, one of its most important features will become apparent: it is the logical bridge to what some scientists believe will be our ultimate energy carrier—gaseous hydrogen produced from solar energy and other renewable resources. Because the two fuels are so similar in their chemical composition—hydrogen can be thought of as methane without the carbon—and in the infrastructure they require, the transition could be a relatively smooth one. (See Chapter 13.) Just as the world shifted early in this century from solid fuels to liquid ones, so might a shift from liquids to gases be under way today—thereby increasing the efficiency and cleanliness of the overall energy system.[47]

6

Winds of Change

By most standards, Denmark is an unremarkable country. With a population of 5 million—less than some cities—and Europe's eleventh largest economy, the nation is hardly an economic or political powerhouse. In energy terms, it is also a bit player, with limited supplies of oil and natural gas in the nearby North Sea. Like many oil-dependent countries, Denmark received a severe shock in the seventies, when energy prices shot up and the country's economy was thrown into a recession. But while other nations responded to the oil crisis by investing in nuclear breeder reactors or synthetic fuels, Denmark took what seemed to be a step backward, turning to an energy source that had been central to its agricultural economy and that of the rest of northern Europe from the twelfth into the early part of the twentieth century—wind power.

Unlike the situation in most other industrial nations, Denmark's search for energy alternatives was not a top-down matter. With a tradition of political decentralization and locally owned farm cooperatives, Danes insisted on a more self-reliant, participatory approach. After extensive public discussions, the country decided to base its energy future on indigenous resources and its own technological ingenuity by focusing on wind power and biomass energy (see Chapter 9). In both cases, Denmark was able to rely on past experience, including the fact that Danish engineers had developed the world's first electricity-generating wind turbine in the 1890s and had continued to advance the technologies as late as the fifties, when wind power was abandoned in favor of cheap oil.[1]

The Danish wind power revival in the seventies began with the scattered efforts of individual inventors who relied on earlier wind machine designs and such off-the-shelf components as old truck transmissions. The country's farm machinery manufacturers quickly joined the effort, realizing that wind turbines would require many components similar to their own. Soon a number of rather primitive wind machines began to appear in the Danish countryside, sparking strong public support and spurring the government energy research laboratory at Riso, which in the past had devoted nearly all its efforts to nuclear power, to join in. Building on earlier work, it took Danish engineers only a few years to develop the technology that today dominates the global wind power industry. The basic concept is a turbine with three fiberglass blades, attached to a steel lattice or tubular tower, similar to those used for power lines. The spinning blades are connected to a standard gear box and generator.[2]

Government researchers in other countries developed more innovative technologies, but they had less success in copying a far more significant element of the Danish program: the active, grassroots involvement of the country's rural population. Starting in the early eighties, a series of wind power cooperatives formed, allowing groups of people to purchase a turbine and lease a site for it, often on the property of a co-op member. Members worked hard to resolve early siting and noise problems and to ensure that wind turbines were adapted to the Danish landscape. The national government played a crucial role by paying 30 percent of the cost in the first nine years, and by requiring utilities to purchase the electricity generated at a fair price.[3]

The results are now visible throughout the Danish countryside. Singly and in small clusters, thousands of sleek, white turbines, 20–30 meters in diameter, stand up like large airplane propellers against the rolling green landscape. Altogether, Denmark had roughly 3,600 wind turbines in operation by 1994, with a capacity of 500 megawatts. This makes Denmark the world's second largest user of wind power, which provided more than 3 percent of the country's electricity in 1994—a figure that is projected to reach 10 percent in 2005.[4]

Danish wind turbines have a well-deserved worldwide reputation for sturdy reliability: turbines designed and built in Denmark are now found on the giant wind farms of California and in smaller groupings in Wales, Spain, India, China, and many other countries. Denmark has effectively led the way to the first intermittent renewable energy technology to be integrated in sizable numbers into the world's electric power systems.

This traditional energy source got a second wind in California in the early eighties. In a political and eco-

nomic climate that could not be more different from
Denmark's, California's pioneering wind power developments were propelled by an odd mixture of environmental activists, venture capitalists, and profit-seeking
dentists and Hollywood moguls. The "wind rush" followed the 1978 publication of a state government study
that unexpectedly found that California had a large
wind energy potential (overturning the conclusions of a
federal study that had been based on data from airports,
which are sited to avoid wind). The report identified
three large, windy passes in California's Coast Range—
at Altamont Pass near San Francisco, in the Tehachapi
Mountains north of Los Angeles, and in the San Gorgonio Pass near Palm Springs.[5]

The study virtually coincided with the adoption of
state and federal tax credits intended to spur wind
power development and with the passage of the federal
Public Utility Regulatory Policies Act, which required
electric utilities to purchase renewably generated electricity at the "avoided cost" of power from conventional
sources. At the time, the Iranian Revolution was driving
the prices of oil and natural gas up, and Californian utilities were building four multibillion-dollar nuclear power
plants, all of them well overbudget. This raised the electricity price available to independent wind producers,
and led to an investment rush similar to the one that
followed the discovery of gold in the Sierra Nevada in
1848.

Between 1982 and 1992, nearly 15,000 wind turbines
were installed in California, most of them before the
federal tax credits expired and electricity markets in the
state became glutted in the mid-eighties. More than 95
percent of these wind farms are in the three large passes
just mentioned, with the rest in similar, smaller passes.[6]

Although the California wind industry peaked early in terms of annual installations, it continued to mature in the late eighties and early nineties, increasing the reliability of the machines and squeezing more and more power out of each turbine. The early "tax farmers," who often used tax breaks to make money on failed wind farms, have been replaced by major brokerage firms and institutional investors, including insurance companies such as Aetna and John Hancock. By the mid-eighties, anyone driving east out of the San Francisco Bay Area on the I-580 freeway was greeted by the sight of thousands of white wind turbines spinning against the brown rolling hills. California's wind farms generated nearly 3 billion kilowatt-hours in 1993, 1.2 percent of the state's electricity, or enough to meet the residential power needs of San Francisco.[7]

When the California wind rush began, developers hoped to use giant turbines with the latest in aerospace technology, which were being built under government contract by companies such as Boeing and Westinghouse. Based on engineering studies that showed large wind machines would be more economical than smaller ones, these sleek and elegant turbines had blades longer than a Boeing 747's wings. Yet not one of these machines proved economical or reliable enough for commercial use, and after several costly experimental installations, the early wind developers turned to smaller machines, including a mix of two- and three-bladed models as well as several vertical axis turbines that looked like large eggbeaters.[8]

The surge in development led a number of small manufacturers to rush their turbines to market; though they were able to get the industry going, few of the early wind machines stood the test of time. In some cases, whole wind farms with hundreds of machines had to be abandoned when they proved too expensive to repair. A few of these wind power "graveyards" are still standing in the mid-nineties. The Danish wind turbine industry came to California's rescue, supplying some 7,500 turbines to the state's developers (twice as many as the Danes installed in Denmark itself).[9]

Since then, the combined efforts of Danish and U.S. manufacturers, with continuing government support, have led to steady improvements in wind generating technology. According to one study, the average capacity factor of California wind turbines—the percentage of their annual power potential they generate—rose from 13 in 1987 to 24 in 1990. These machines are now estimated to be "available" to operate 95 percent of the time, which is better than most fossil fuel plants.[10]

As the technology has developed, the cost of capturing the wind's energy has fallen precipitously. In the early eighties, wind machines typically cost $3,000 per kilowatt and produced electricity for more than 20¢ a kilowatt-hour (1993 dollars). By the late eighties, the machines were larger and more efficient and their capital cost, including installation, had fallen to about $1,000–1,200 per kilowatt. At an average annual wind speed of 5.8 meters per second (13 miles per hour) and a maintenance cost of a penny per kilowatt-hour, this yields an average generating cost of about 7¢ per kilowatt-hour for wind turbines installed in the United States in the early nineties, compared with 4–6¢ for new power plants fueled by natural gas or coal.[11]

More than a dozen American and European companies are pursuing advanced wind technologies, many with government assistance, that are believed capable of closing the remaining cost gap with fossil fuel plants.

The machines now entering the market generate 300–750 kilowatts per turbine rather than the 100-kilowatt average of the late eighties. They have lighter and more aerodynamic blades, improved rotor-hub connections and drive trains, new aerodynamic and electronic blade controls, and more advanced power electronics, including some that operate at variable speeds, which allows the turbines to operate more efficiently at a range of wind speeds. The new designs are less expensive and can be deployed in more moderate wind regimes. In fact, wind developers using the new technology have signed contracts to sell wind-generated electricity at less than 5¢ per kilowatt-hour.[12]

As wind power has gained acceptance as a reliable, economical source of power, a number of U.S. utilities have begun to incorporate it into their plans. California, for example, is poised to end its wind power hiatus in the mid-nineties, with some 585 megawatts' worth of projects bidding successfully in a power auction conducted by Pacific Gas and Electric and San Diego Gas and Electric in late 1993. Sizable wind power projects are also being built or planned in Iowa, Maine, Minnesota, Montana, New York, Oregon, Texas, Vermont, Washington, Wisconsin, and Wyoming, many of them being developed by the Kenetech Corporation, the largest U.S. wind power developer. In 1993, the American Wind Energy Association set a goal of pushing U.S. wind power capacity to 10,000 megawatts by 2000, a target that will require accelerated development in the late nineties.[13]

Despite the surge in wind power in the United States, the nineties are beginning to shape up as the decade of Europe in this field. Europe's northern plains, bordering the North and Baltic Seas, have a particularly large wind power potential, extending as far east as Poland. And Mediterranean winds are strong in southern Europe. Prompted by growing environmental concerns, several European governments stepped up their efforts to promote carbon- and sulfur-free renewable energy sources in the late eighties, including increased R&D support and additional subsidies.

In northern Germany, where wind power is growing rapidly, a relatively high power purchase price and a generous government subsidy allow private wind power developers to receive 13¢ per kilowatt-hour for their electricity. In the Netherlands, an integrated wind power development program is a prominent part of the country's National Environmental Policy Plan. And in the United Kingdom, the newly privatized power industry is required to reserve a portion of new power contracts for "non-fossil" sources, an obligation originally intended for nuclear power but now used to promote wind energy.[14]

While Denmark continues to add to its wind power installations, its total capacity will soon be rivalled by Germany, the Netherlands, and the United Kingdom, each of which has set goals of developing at least 1,000 megawatts of wind-generating capacity by 2005. Italy and Spain are moving forward as well, though more slowly. All this activity has created a lot of business, and by the mid-nineties the European wind industry—more than a dozen active manufacturers—was booming. Altogether, the European Union has plans to install 4,000 megawatts of wind power capacity by 2000 and 8,000 megawatts by 2005—more than six times the 1993 total.[15]

The wind rush has also spread to other parts of the globe. In 1993, the Ukrainian electric utility Krimenergo and Kenetech Corporation of San Francisco signed a deal to install 500 megawatts of wind turbines in the Crimea Peninsula. In India, the World Bank approved a $175-million loan for renewable energy in 1992, driving up land values in two wind-rich areas in Tamil Nadu to between 8 and 20 times what they were five years earlier. Mexico has identified a region south of Mexico City, called La Ventosa (the windy area), with several thousand megawatts of wind power potential. Argentina, which has some of the world's strongest winds, is working with Danish companies to deploy 500 megawatts of wind generators and a domestic industry to manufacture them. Other sizable wind power projects are under way in China, New Zealand, and several other countries.[16]

The world had roughly 20,000 wind turbines in operation by the end of 1993, producing about 3,000 megawatts of electricity (90 percent of it in California and Denmark), 30 times as much as a decade earlier. (See Figure 6–1.) Although wind power still provides less than 0.1 percent of the world's electricity, it is fast becoming a proven power option, considered reliable enough for routine use by electric utilities. According to the Electric Power Research Institute, wind power offers utilities pollution-free electricity that is nearly cost-competitive with today's conventional sources. Forecasts suggest that wind power costs should fall to less than $800 per kilowatt (4¢ per kilowatt-hour, including operating costs) by the end of the decade, down from about $1,000–1,200 per kilowatt in 1993, and perhaps one day to $600 per kilowatt or 3¢ per kilowatt-hour. Much of the cost reduction will be driven by the economics of mass production. These figures suggest that wind power soon may be one of the least expensive sources of electricity.[17]

The biggest outstanding concern among utilities is

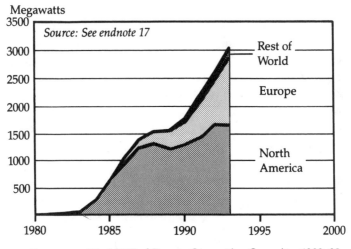

FIGURE 6-1. *World Wind Energy Generating Capacity, 1980–93*

how to integrate intermittently operating wind turbines into their grid systems. The early wind farms have caused few such problems, thanks in part to the evolution of new electronic controls, but if the wind is to supply 20 percent or more of a region's electricity, some adjustments are needed.

Most of today's power systems operate with a combination of "baseload" power plants that run most of the time (generally fueled by coal or nuclear power), "intermediate" plants that are turned off at night, and "peaking" units that operate only when demand is highest—usually gas turbines or hydro plants. Wind power does not fit neatly into any of these roles, and because of its variability, the rest of the power system has to be managed so as to ensure adequate backup power. (See Chapter 12.) But engineers say that the challenge is not qualitatively different from one that utilities mastered long ago: meeting the rapidly fluctuating demands of customers.[18]

By carefully studying wind patterns and integrating wind power into the overall system, problems can be minimized. In regions with large hydro dams, such as the northwestern United States or Brazil, no additional backup is required. But in other areas, heavy reliance on wind power may require backup gas turbines or even compressed air or pumped hydro storage. Depending on the local geography and geology, storage technologies will vary in cost, but engineers are optimistic that the devices will become more cost-effective in the years ahead. Gas turbine technology is already sufficiently inexpensive to make it an attractive standby generator. With these and other technologies, large amounts of wind power can be integrated into power systems without in any way threatening reliability. In fact, in some countries the wind power share could eventually go as high as 30–50 percent even without storage (though with an additional cost of roughly 10 percent).[19]

Questions nevertheless remain about the economics of intermittent wind power. Electric utilities value new power plants based on their ability to offset fuel and operating costs at other plants, as well as their "capacity value," the measure of their ability to supply power when it is most needed. Experience so far shows that in some regions peak winds coincide nicely with peak power demand, but in others they do not. In northern California's Altamont Pass, for example, there is a good but not a perfect match. Peak power demand comes between 2:00 p.m. and 8:00 p.m. on summer afternoons, but the highest winds occur between 5:00 p.m. and midnight.[20]

The worth of a given quantity of wind power will

therefore vary widely, but on most electrical systems, wind farms do increase the ability to provide reliable power. Recent studies suggest that well-designed and economically optimized wind farms in the Great Plains of the United States will have a capacity factor of 30–50 percent, compared with 75–81 percent for a typical coal plant and 68 percent for a nuclear plant. Depending on how well matched the wind is to power needs, these figures suggest that a wind farm can have a capacity value that is up to 70 percent of that of a coal plant. Based on this, and on their extremely low operating costs, it appears that wind farms costing $700–800 per kilowatt—expected to be available by the late nineties—will be fully competitive with fossil-fuel-based plants. In later years, they could be less expensive than coal.[21]

Even excluding environmentally sensitive areas, the global wind energy potential is roughly five times current global electricity use. Since the power available from wind rises with the cube of the wind speed, most of the development will occur in particularly windy areas. In the United States, where detailed surveys have been conducted, it appears that wind turbines installed on 0.6 percent of the land area of the 48 contiguous states—mainly in the Great Plains—could meet 20 percent of current U.S. power needs. Indeed, conservative resource estimates that exclude large environmentally sensitive areas show that three U.S. states—North and South Dakota and Texas—could in theory supply all the country's electricity. Although no one expects such a scheme to actually be implemented, it demonstrates the potential for wind power to become a major component of North America's power systems. (See Table 6–1.)[22]

Not all nations have such an abundance of wind potential, but many do. Among the other countries that

TABLE 6-1. *Wind Electricity Generation and Potential,*
Selected Countries

Country	1993 Wind Generation	Potential[1]	Potential as Multiple of Generation
	(terawatt-hours)		
United States	2.9	10,777	3,700
Denmark	1.1	38	36
Netherlands	0.1	16	128
United Kingdom	0.2	760	3,600

[1]Potential excludes land not available for wind development due to environmental, urban, and other conflicting uses, as well as offshore resources; assumptions for determining potential vary slightly by country.

SOURCE: See endnote 22.

could in all likelihood match the United States by supplying (in theory) all their electricity with the wind are Argentina, Canada, Chile, China, Russia, and the United Kingdom. Others, such as Egypt, India, Mexico, Tunisia, and South Africa, should easily be able to push their reliance on wind power to 20 percent or more. Europe as a whole could obtain between 7 and 26 percent of its power from the wind, depending on how much land is excluded for environmental and aesthetic concerns. In addition, many regions with limited wind power potential may find that the limits end at the water's edge: New England, the United Kingdom, and Poland are among the areas that could obtain most of their power from wind farms located on offshore platforms in shallow seas. At least 20 small subtropical island countries have nearly constant trade winds that could meet a large share of their electricity needs. Since many of these nations now depend on expensive diesel generators, wind power is an attractive option.[23]

Relying on wind power as a major energy source will inevitably generate land use conflicts. On the plus side, while it is true that wind farms would "occupy" large areas (1 percent of the land of the United States to supply one third of the country's electricity), most of it would be land where few people or wildlife reside. Moreover, wind machines occupy land mainly in a visual sense. The area surrounding the turbines can be used as before—usually for grazing animals or raising crops—and provide farmers with supplementary income. In Wyoming, for instance, a hectare of rangeland that sells for $100 could yield more than $25,000 worth of electricity annually. In many windy regions, harnessing the wind might enhance land values by acting as a windbreak and reducing susceptibility to erosion.[24]

Still, the accelerating pace of wind power development in the nineties has led to a few environmental controversies. From Rattlesnake Ridge in the state of Washington to the rolling hills of Wales, local groups have protested wind power as a visual blight on the landscape. Some developers have ignored the participatory approach pioneered in Denmark and sought to dismiss or bully the opposition—a practice they have later come to regret. Ironically, Rattlesnake Ridge is adjacent to the U.S. government's nuclear weapons facilities in Richland, where some of the wind power protest has been led by resentful nuclear advocates. Still, many opponents have raised valid concerns, and unless these are addressed seriously and democratically, with environmentally sensitive areas set aside as a result, wind power is unlikely to become a major energy source. On the other

hand, experience suggests that in many regions the aesthetic appeal of wind farms has grown over time. (It is
worth remembering the Golden Gate Bridge was originally opposed on aesthetic grounds as well.)[25]

An even more serious problem gained attention in the
early nineties: the potential of wind turbines in some
areas to kill birds, including rare raptors and other en-
dangered species. This was first identified in northern
California's Altamont Pass, where a number of golden
eagles have been killed; dead birds have also been found
in southern California's San Gorgonio Pass and the
European Union–supported wind farm in Tarifa, Spain,
near the Strait of Gibraltar. According to ornithologists,
some bird species are attracted to the same windy areas
that are good for wind turbines, but additional research
will be needed to discover how widespread and serious
these problems are.[26]

Already, officials at the National Audubon Society
and the U.S. Fish and Wildlife Service have called for a
moratorium on wind power development in areas of the
United States with heavy raptor populations, and in
Europe, the European Union issued a similar call. Experts hope that means can be found to scare birds away
from the wind farms, and believe that as wind power
development moves from mountain passes to the open
plains, there will be less conflict with birds.[27]

The other constraint to heavy reliance on wind power
is the long distances that separate some of the world's
large wind resources from the major population and industrial centers where most energy is used. This problem is seen clearly in the United States, where nearly 90
percent of the country's huge wind resource is in the
Great Plains, more than 1,000 kilometers from Chicago
and 2,000 kilometers from New York, Atlanta, and Los

Angeles. Long transmission lines will be needed to carry wind energy to where it is needed; existing lines are often inadequate or in the wrong location.[28]

Building the new lines is feasible, but will take time and money. Moreover, developers face a chicken-and-egg problem: until large power projects are under way, there is no incentive to build a long transmission line, but no one wants to invest in a wind farm unless they are sure they can get the power to market. In addition, new transmission lines face siting problems that are far more severe than those faced by wind farms, as a result of growing controversy over the potential health effects of electromagnetic radiation from high-voltage lines. Over time, most of these obstacles can be overcome. Direct current lines can avoid the radiation problem, and in sensitive areas, the lines can be placed underground. Also, electronic technologies are under development that will increase the carrying capacity of large power lines, and reduce line losses as well.[29]

The economics of remote wind power also appear favorable. A 2,000-kilometer, 2,000-megawatt transmission line would cost roughly $1.5 billion, which would add only about a penny per kilowatt-hour to the delivered cost of wind energy. The challenge is to put together consortia of electric utilities and private power developers to assemble these large projects. In the long run, large wind farms developed in remote areas may also be used to produce hydrogen during off-peak periods, which would then be fed into a hydrogen pipeline and storage system. (See Chapter 13.)[30]

The big question for the late nineties and beyond is how quickly the global wind power market will develop. Although the rapid advance of the technology and the spread of commercial projects provide grounds for optimism, continuing, targeted government support will be needed in order to make this a self-sustaining industry. In the past, much of the governmental effort on behalf of wind power has focused on research and development; while R&D remains important, commercialization is now the highest priority.

The key to wind power development is the stance of utilities, most of which are unfamiliar with the technology and resist anything new. Opening the grid to independent wind developers, as California and Denmark have done, and giving them attractive but fair prices for the power they provide is the key to boosting the industry and attracting the capital needed to develop the next generation of wind turbines. (See Chapter 12.) In addition, small, declining subsidies or tax breaks (of the sort now used in Germany) can spur wind power development. Finally, governments can help utilities form consortia that commit to building large numbers of turbines, thereby allowing manufacturers to scale up production and bring down the price.

Several countries have adopted some or all of these measures in recent years as a way of encouraging the development of less-polluting power sources. If such policies continue to spread, wind power is likely to emerge as a major global energy resource soon after the turn of the century.

7

Heat from the Sun

If necessity is the mother of invention, then Israel is well positioned to help lead an energy revolution. Despite being just a short distance from the world's largest oil reserves, the Jewish state has no significant fossil fuel resources of its own and, for obvious political reasons, does little trading with its energy-rich neighbors. The product of this deprivation can be seen on virtually every Israeli rooftop: nearly a million solar collectors, which in 1994 provided hot water for an astonishing 83 percent of Israel's homes. Even in the heart of the Middle East, solar energy is the most abundant energy source. One day it may also be the most widely used.[1]

Solar water heating actually got its modern start in Palestine in 1940, nearly a decade before the Jewish state was founded. With much of the world at war, Levi

and Rina Yissar struggled to find a way to provide warm
bath water for their infant son. Rina seized on the idea of
painting an old tank black, filling it with water, and leaving it out in the desert sun to warm. Not only did this
provide her son with dozens of warm baths, it sparked
the imagination of her husband. Levi Yissar became
convinced that solar energy might be a viable way of
heating water in Israel, and so he began a search that
eventually took him to Florida and California. Modifying existing designs, Yissar came up with a box-like,
roof-mounted solar collector made of copper and glass
that he thought would be attractive to his neighbors.[2]

Despite the skepticism of many Israelis, Yissar assembled and distributed several solar hot water systems,
demonstrating that solar heating was reliable and would
pay back the initial investment in just two years. By
1953 he had established a full-fledged solar collector
manufacturing company, and he demonstrated his business acumen by selling one of his first systems to Israel's
founder, David Ben Gurion. The rest, as they say, is
history. While other countries were left flatfooted by the
1973 oil crisis, Israel responded by increasing its support of solar hot water and spurring the development of
the solar industry. The government eventually required
all residential buildings up to nine stories high to use
solar energy for water heating.[3]

Today, Israel has the world's greatest density of solar
collectors. Some 900,000 solar hot water systems were
in place by 1994, providing business for 30 solar companies. These successes in turn led Israel to develop more-
advanced means of harnessing the sun's energy, including salt-laden solar ponds and mirrored solar troughs,
each of which can provide temperatures high enough to
boil water and generate electricity. Today, Israel also

has one of the highest concentrations of solar scientists
and engineers, and the fruit of their labors can be seen
around the world. While its centralized energy institutions have so far constrained Israel from replicating the
solar success story with other energy technologies, the
nation has already provided the world with an important
lesson: solar energy can make a mockery of the "energy
supply dilemma," even in countries without any fossil
fuels. But this energy will only be harnessed if traditional
priorities are reordered and if governments free up the
human ingenuity critical to its widespread use.[4]

★ ★ ★ ★

It took the Industrial Revolution to break humans' almost total dependence on solar energy. Our ancestors
relied on sunlight to warm their dwellings and dry their
crops, and on indirect forms of solar energy, from wood-
fueled cook fires to ox-drawn ploughs and water-driven
grinding mills, to meet important needs. (Photosynthesis uses sunlight to produce the complex biological molecules found in living plants.) Early humans simply took
advantage of the fact that each day a tiny portion of the
solar energy that strikes the earth's surface is converted
into wind, rain, and plant matter. Even today, some of
these stored forms of solar energy are the mainstays of
agricultural economies in much of the developing world.

Direct sunshine is the most abundant energy source
of all. Some 5.4 million exajoules of solar energy strike
the earth's upper atmosphere each year, one third of
which is reflected back into space, with another 18 percent absorbed by the atmosphere—much of it converted into wind. This leaves 2.5 million exajoules that
reach the earth's surface, more than 6,000 times the
amount of energy used by all human beings worldwide

in 1990. Stated otherwise, the entire exploitable resource of fossil fuels in place when civilization began—oil, coal, natural gas, and tar sands—is equivalent to less than 30 days of sunshine striking the earth.[5]

Most of this energy is quickly transformed into heat that is absorbed by the oceans and the atmosphere. (Without solar energy, our homes would be -240 degrees Celsius when we turn on the furnace.) While heat is a relatively "disorganized" form of energy that lacks the versatility of stored chemical fuels or electricity, figures for the United States show that 40 percent of the energy delivered to the country's factories, motor vehicles, and buildings is used for heating and cooling, most of it at relatively modest temperatures of less than 100 degrees Celsius. Moreover, a host of new technologies now in use or under development allow solar energy to be concentrated so as to provide high-temperature heat, mechanical power, or electricity. Many scientists believe that this abundant energy source has the potential in the next century to become the prime mover of the global economy. To reach this point, an economical way to convert this "low-grade" energy to usable forms will be needed, along with the means to store it.[6]

In the seventies, solar energy suddenly was transformed from a scientific curiosity into a cultural movement. Social activists saw it as a symbolic and practical alternative to fossil fuels and nuclear power. Thousands of tinkerers built passive solar homes and wore small buttons blazoned with the slogan "Solar Si! Nuclear No!" The movement crested on May 3, 1978, when hundreds of thousands attended solar demonstrations, lectures, and tours in the United States on Sun Day. Although solar energy as a political movement soon faded, the broader legacy continued as governments stepped up research programs and provided a variety of tax breaks and other subsidies for solar development, leading a host of new companies to enter the business.[7]

So far, efforts to harness solar energy have been most successful when they have concentrated on the simplest and most economical of applications. Since three fourths of the energy used in buildings goes to space and water heating, space cooling, and refrigeration—services that can be provided with low-temperature heat—a convenient market for solar energy exists. Most simply, the sun's rays can be harnessed by modifying building construction, design, or orientation to keep a building warm in winter and cool in summer, with only minimal use of other fuels. (See Chapter 11.) In addition, as Israel has shown, collectors installed on roofs or along the facades of buildings can take in solar energy that is used for water or space heating.[8]

First developed by a Swiss scientist in the eighteenth century, most "flat-plate" solar collectors consist of a glass-topped box with a copper absorber plate and small copper tubes through which the heating water flows from bottom to top—usually connected to an indoor storage tank. Solar hot water systems gained popularity in California, Florida, and Australia at the turn of the century, though they faded with the advent of cheap oil and natural gas. In the late seventies, solar water heating made a strong comeback, propelled by tax credits and the desire to achieve more energy independence. In the United States alone, dozens of companies installed more than 1.5 million solar heating systems during this period. Although the market collapsed in the mid-eighties with the expiration of the tax credits, the solar industry again mounted a comeback in the early nineties.[9]

Japan, which until recently has suffered from high fuel

prices and the lack of a natural gas distribution system, also expanded its reliance on solar water heating, and is now the world leader. Some 4.5 million Japanese buildings had solar hot water systems in 1992, though sales have declined by two thirds since the mid-eighties. In Europe, solar water heating was until recently quite rare, but environmental concerns in the early nineties helped spur government promotion efforts. In the Netherlands, for example, some 10,000 solar hot water heaters had been installed by more than 10 companies by the end of 1993, spurred by subsidies that are due to be phased out as the industry becomes stronger in the late ninteties.[10]

Solar water heating has also caught on in a number of developing countries, often with little government support. Solar energy's popularity is frequently heightened by the lack of natural gas, forcing consumers to rely on electricity—which is far more expensive—for their water heating. Residents of Botswana's capital, Gaborone, have purchased and installed more than 3,000 solar water heaters, displacing nearly 15 percent of the country's residential electricity demand. Some 30,000 solar hot water heaters have been installed in Colombia, 17,-000 in Kenya, and nearly 1,000 in Malawi. And in Jordan, 26 percent of the country's households use solar hot water systems.[11]

The cost of flat-plate collectors fell 30 percent between 1980 and 1990 as a result of improved designs and manufacturing techniques. According to the American Solar Energy Society, heating water with the sun can now cost as little as $11 per gigajoule. This is about half the average residential price of electricity in the United States, close to the price of bottled propane gas, and twice the price of piped natural gas. Still, until the market expands and manufacturers are able to mass-produce collectors, costs will remain relatively high. In the meantime, the economics of installing a solar hot water system have to be judged on a case-by-case basis, with the main variables being climate, the cost of competing fuels, and interest rates. In the future, rising gas prices and falling solar collector costs may close the price gap.[12]

Somewhat surprisingly, sunlight can also be used to cool buildings, either through natural ventilation techniques, absorption cooling devices that run on a normal refrigerator cycle by condensing and evaporating a refrigerant fluid, or desiccant cooling systems, which use a drying agent to adsorb water vapor, reduce humidity, and cool the air through evaporation. These processes are not as efficient as the simple harnessing of solar heat, a fact that is partially compensated by the concurrence between periods of peak sunlight and peak cooling needs. Solar cooling could help meet one of the developing world's fastest growing energy needs—air-conditioning—which is being fueled by the construction of office and apartment buildings in countries with tropical climates.

A prototype solar absorption system that was installed in a small office building in California's scorching Central Valley in 1990 provides more than 80 percent of one building's cooling, with the remainder derived from off-peak electricity used to chill water at night. The same array of rooftop solar panels provides half the building's space heating in winter and more than 80 percent of its hot water. A natural gas cooling system under development works on the same principle, raising the possibility of a "hybrid" that uses the solar collectors as a fuel-saving supplement to a gas-driven cooling system. The largest initial market is likely to be commercial buildings, but solar cooling may one day be broadly cost-effective in sunny climates.[13]

In order to expand the solar collector market and help manufacturers to scale up production, a few electric utilities have begun to help customers install solar water heaters. The Sacramento Municipal Utility District (SMUD) in California is financing the conversion of electric water heaters, chalking up more than 1,800 systems by early 1994. In 1993, the utility offered a performance-based rebate up to $863 on systems that cost less than $3,000. Customers pay the remainder through their bills over 10 years. According to SMUD, the monthly fee is less than its average bill for electrically heated water. At the same time, the utility began a tough certification and inspection program to ensure reliability. SMUD's goal is to have more than 20,000 solar systems in place by 1999, covering half the homes in the area that heat water with electricity.[14]

Although solar heating has achieved much of its recent popularity on middle-class roofs, it has a much more critical role to play outside the tiny huts that house hundreds of millions of poor people in the developing world. For them, cooking is often the largest use of energy, and widespread deforestation has reduced the availability of fuelwood. Fortuitously, most of the world's poor live in sunny climates near the equator, raising the prospect of cooking with the sun. Unfortunately, many of the early efforts to develop solar cookers focused on elaborately engineered stoves involving mirrored parabolic dishes. Even those that worked were far too expensive for the intended rural market.

In 1976, Arizonans Barbara Kerr and Sherry Cole rewrote the book on solar cooking when they developed a solar cooker that consisted of an insulated box with a transparent, removable top. By adding a single reflector to increase the sunlight entering the box, interior temperatures can top 100 degrees Celsius, more than enough to cook food in the three to four pots that fit inside. The boxes are relatively easy to build, and the materials to construct them—a piece of glass, aluminum foil, and a bit of glue and cardboard—cost as little as $12 per cooker. If the sun is strong, rice can be cooked in two hours, corn in three to four, and beans in five to six. Moreover, solar box cookers are easy to use, require little tending, and allow people—mainly women—to spend less time collecting firewood and bending over smoky cookstoves. The same devices can improve rural sanitation since they heat water sufficiently to kill most coliform bacteria, amoebas, giardia, and the rotavirus that causes the death of more than 2 million infants a year.[15]

Efforts to promote solar box cookers have been under way in a host of countries since the mid-eighties, much of it led by Bob Metcalf, a professor of biology at California State University, Sacramento, who built some of the first cookers in his backyard in 1978, and who still cooks with one. With the support of several nonprofit groups and private foundations, Metcalf has traveled extensively in Africa and Latin America, demonstrating his cookers to people and encouraging them to set up local enterprises to build and maintain them. In Guatemala's El Progreso region, more than 50 families were using box cookers on a regular basis in 1988, reducing their sizable expenditures on wood cooking fuel. In the treeless highlands of Bolivia, Freedom from Hunger has trained 50 local promoters in solar cooker construction and use, and similar efforts have been mounted in Djibouti, Mexico, Zimbabwe, and other countries.[16]

Learning from the mistakes of other appropriate technology missionaries, the prophets of solar cooking focused on designing a cooker that meets the needs of people, rather than expecting people to adapt to the demands of the cooker. Still, solar cooking has often been a hard sell. Each culture has its own unique customs and rituals that govern the cooking process, and many people prefer the taste and texture of food roasted over a hot fire to that of food baked in a slow-burn oven. As a result, many villagers initially have rejected solar cooking as incompatible with their traditions.

Nevertheless, there are promising signs that where wood fuel is in short supply, solar cookers are catching on. In India, for example, local self-help groups and the government have promoted a modified box cooker that has been distributed to more than 200,000 families. Solar cooking has also caught on in parts of China (where parabolic dishes are promoted) and in Pakistan, where in the early nineties some 10,000 solar box cookers were being used by Afghan refugees in fuelwood-scarce areas such as Peshawar and Quetta. If promoters continue their steady, patient efforts to introduce people to solar box cookers, and if local entrepreneurs develop the ability to manufacture, distribute, and repair the systems, solar cooking may one day become a critical part of the energy economy in many regions.[17]

* * * *

Although simple solar heating technologies have a major role to play, the key to more extensive reliance on solar energy in the twenty-first century is developing a cost-effective means of concentrating sunlight so that some of the immense amounts of energy that strikes the earth each day can be converted into mechanical energy, electricity, or chemical fuels—the energy forms on which much of today's global economy must run. Since the mid-seventies, numerous alternatives have been explored both by governments and by private researchers, often with the aim of converting sunlight to electricity, a particularly expensive and versatile form of energy.

Government energy planners in the seventies, accustomed to the huge size and economies of scale of the electric power industry, commissioned engineering studies that reached a predictable conclusion: large, centralized solar power plants were the answer—analogous to the giant coal and nuclear facilities that the same consultants were urging. Huge "power towers" and "solar furnaces" were called for in these reports, and some even suggested a series of massive orbiting solar collectors that would beam intense microwave energy back to earth. Although the idea of an orbiting power station has largely been abandoned as economically impractical, several monuments to this vision of solar energy on a massive scale were built in the eighties in France, Japan, Italy, Spain, Ukraine, and the United States.

Among the large solar generators that got the most attention was Solar One, a 77-meter-high, tower-mounted receiver built in southern California's Mojave Desert in the early eighties. It is surrounded by seven hectares of mirrors (called heliostats) that reflect sunlight onto the top of the tower, where water is flashed into steam used to power a 10-megawatt generator on the ground. Although Solar One turned out to be an expensive and unreliable power plant, the engineers who designed it are working on a new design, dubbed Solar Two, that includes a molten salt storage system. Engineers project that a 100-megawatt version may have a generating cost as low as 8¢ per kilowatt-hour.[18]

Solar Two looks great on paper, and it is expected to provide steady baseload electricity as well as late afternoon peaking capacity, but the future of all the central solar generators is in doubt. They are expensive to build, their very scale escalates financial risks—as with nuclear power—and their massive height (in excess of 200 meters) may attract opposition.

In focusing so narrowly on the most centralized of solar power technologies, government and utility engineers missed two simple but crucial points: Solar energy's great advantage over other energy sources is that it is reasonably dispersed—providing a better match to human energy use patterns. Second, and perhaps more important, is the fact that it is far easier to develop and commercialize a technology that can be installed in small, modular units that are gradually improved over time. The power tower depends on the vagaries of government funding and faces a continuing threat that the next technical setback may be its last.

To find a more practical means of concentrating the sun's energy, scientists in effect took another step backward. In the 1870s and 1880s, at the height of the Industrial Revolution but before the use of oil had become widespread, French and U.S. scientists had developed an array of solar cookers, steam engines, and electricity generators, all based on a relatively simple concept: a parabolic-shaped solar collector that is coated with a mirrored surface to reflect light coming from different angles onto a single point or line.[19]

Two variants were developed: parabolic "dishes," which resemble satellite television receivers, or trough-shaped parabolic collectors that concentrate sunlight onto a tube rather than a single point. By the turn of the century, clever inventors had built a wide array of solar engines that did everything from running a printing press in Paris to pumping irrigation water in Arizona. Within a few years, however, the era of cheap oil had begun, and most of these efforts to harness solar energy were abandoned.

Seven decades later, in the aftermath of the 1973 oil embargo, parabolic solar dishes and troughs made a sudden comeback. Within a short time, scores of scientists and engineers were at work developing new designs. Though their efforts were not as well funded as the larger power towers, the results have been more promising. Late-twentieth-century inventions such as inexpensive reflective materials, improved heat transfer fluids, more-efficient solar receivers, and electronic tracking devices have greatly improved the effectiveness of a nineteenth-century technology, allowing engineers to reduce the cost of these systems during the eighties.

The most successful effort to commercialize solar trough technology was launched in Israel in 1979 by Arnold Goldman, the inventor of word processing. Goldman designed a system of mirrored troughs 9 feet high and 40 feet long to concentrate the sun's rays onto oil-filled tubes that run parallel to the mirrors along their focus. The troughs are mounted on a device that allows them to follow the arc of the sun, keeping it focused on the collection tube. Goldman's original concept was to mount these collectors on the flat roofs of industrial buildings, providing heat for textile production, food processing, and other industries. Although this application has failed to catch on, it still appears to hold promise, particularly in developing countries that lack natural gas. At least one U.S. investment group was rumored to be ready to market a solar process steam system in the mid-nineties.[20]

Being an entrepreneur, Goldman was attracted to another market in the early eighties: the generous power purchase contracts and solar tax credits being offered in California. Soon his firm, Luz International, had lined up several large contracts to deliver electricity to the Southern California Edison Company, and was at work on a series of solar power generators. In Luz's design, the solar trough heats a working fluid that circulates to a power station where water is superheated into steam, which then powers an electricity-producing turbine. And by burning natural gas as a backup fuel, Luz is able to keep its turbines running even when the sun is not shining.

After gradually fine-tuning the solar field and figuring out how to operate it most efficiently, Luz can now convert 23 percent of incoming sunlight into electricity during peak sunny periods and 14.5 percent on an annual basis. The power plant built at Harper Lake, California, in the Mojave Desert in 1989 is a good example. Spread over 750 hectares, hundreds of rows of gleaming solar collectors produce enough power for about 170,000 homes for as little as 9¢ per kilowatt-hour, which is competitive with generating costs in some regions, and far below the 29¢ cost of Luz's first plant in 1984. Without using natural gas for backup, however, the latest Luz plant can generate electricity for 13¢ per kilowatt-hour, with the potential to cut that figure to less than 10¢ by scaling up production. (See Figure 7–1.)[21]

Between 1984 and 1990, Luz installed nine plants with a capacity of 354 megawatts of generating capacity. In fact, the company shipped enough equipment to California's Mojave Desert in the late eighties to place it among the top 10 Israeli exporters. But Luz's fortunes took a bad turn in 1989, when falling natural gas prices drove the company's power purchase price down at the

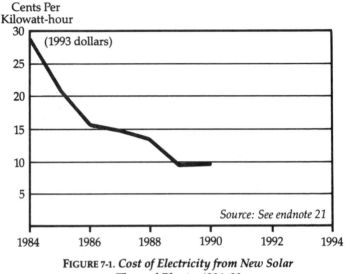

FIGURE 7-1. *Cost of Electricity from New Solar Thermal Plants, 1984–90*

same time the vagaries of a tax-incentive-driven market caught up with it.[22]

Unlike the tax breaks for the U.S. oil and gas industries, which amounted to $8.8 billion in 1989, the federal solar tax credit required annual renewal. In 1989, Congress renewed the credits for 9 months instead of the normal 12, forcing Luz to finish its 1990 project 3 months earlier than planned and driving up costs. The final blow came a few months later when a faulty analysis led California's finance department to temporarily revoke the company's property tax exemption, leading nervous investors—who had already put more than $1.25 billion into Luz—to pull the plug.[23]

Still, even after the company's demise, Luz's nine plants continued to churn out power in the Southern California desert. Belgian investors purchased the rights to the technology from Luz's creditors, calling the reorganized company (which is still based in Israel) Solel. As of early 1994, Solel was focusing on R&D, working to raise the efficiency and improve the storage of its trough systems, while rumors were rampant of negotiations under way to build one or more 200-megawatt solar power plants, possibly in Brazil, India, Israel, or Morocco.[24]

Other research teams also were developing solar troughs in the early nineties, including an innovative design by David Mills of the University of Sydney. Tests reveal that improved collector surfaces, polar axis tracking, and an improved trough design can increase sunlight collection by nearly one fourth. By also adding vacuum insulation to the heat-carrying pipes, Mills estimates that the annual efficiency of the system may reach 20 percent, with peak efficiency between 25 and 30 percent.[25]

The new design also has the ability to run for eight hours without sunlight, by storing heat in an inexpensive bed of rocks. Such a plant could provide "baseload" power for around 6¢ a kilowatt-hour, and would also be able to do something that most fossil-fueled plants cannot: economically follow the ups and downs of customers' power needs. As Mills puts it: "Despite the tenacious myth that baseload is beautiful, the ideal power plant would produce an output that was as flexible as possible. . .accounting for diurnal variations in demand as well as changes in the availability of other generating plants on the utility grid."[26]

The parabolic dish collector is also mounting a strong comeback. Each dish follows the sun individually, with double-axis tracking that allows sunlight to be focused onto a single point where heat can be either converted directly to electricity or transferred by pipe to a central turbine. Parabolic dishes are generally more thermally efficient than troughs. Moreover, since parabolic dishes (like troughs) are built in moderately sized, standardized units, they allow generating capacity to be added incrementally as needed. And they can attain temperatures three to four times that of trough systems, thus producing higher quality steam and more electricity.[27]

The U.S. Department of Energy expects parabolic dishes to produce power for 5.4¢ a kilowatt-hour early in the next century, and some experts believe that such systems will outperform troughs economically. One system has been designed at the Australian National University, and an initial 2-megawatt plant was being built in 1994 in Australia's remote Northern Territory. The project is being funded by a consortium of electric utilities under pressure to reduce their heavy reliance on coal. Twenty-five dishes will feed steam to a turbine to produce power for the isolated community of Tennant Creek.[28]

Meanwhile, advances in Stirling engines (also known as heat engines) are opening up another use of dish-shaped solar collectors. Mounted at the focal point above the dish, the Stirling engine directly converts heat to electricity, reaching conversion efficiencies as high as 29 percent. These systems can heat homes and factories, but may be particularly practical as a way of providing small amounts of power in remote areas.[29]

Another approach to collecting the sun's energy is to rely on abundant and cheap salt water to collect solar heat. The difference in temperature between warm surface water and deeper water layers can be used to drive a heat engine to generate electricity. When cold water passes warm water in a heat exchanger, the warm water evaporates and drives a low-pressure steam turbine to produce power.

Two variants of this technology have been developed: solar ponds and ocean thermal energy conversion (OTEC). In the first, an artificial lake of salty water developed by Israeli scientists in the seventies was used to generate 5 megawatts of electricity on the shores of the Dead Sea for several years until it was closed in 1989 due to high costs; smaller systems were also built in Australia and California. The low conversion efficiencies of these systems are countered by their low cost per square meter. Still, solar ponds have yet to attract serious commercial interest, and may not have the long-run economic potential of other solar thermal technologies.[30]

The ocean thermal variation of the solar pond concept may have greater promise. Solar energy that strikes the oceans can be captured in a similar manner in regions where surface and deeper waters differ in temperature by at least 20 degrees Celsius. OTEC uses the difference between sun-warmed surface waters in tropical areas and deep, cold water to power a heat engine, and could supply electricity around the clock. A pipe as long as 1,000 meters is used to pump huge volumes of water to the surface, where it is run through a heat exchanger. One variant on the design is to couple a deep-water collection system with an on-shore solar pond, raising the temperature differential and lowering the amount of water that must be moved.

Although large OTEC systems face sizable engineering hurdles, and some analysts have questioned their economics, the value of these systems is enhanced by the fact that they can produce two other valuable commodities: fresh water and fish that can be raised in enclosed pens supported by the deep-water nutrients brought to the surface by an ocean thermal system. A prototype system in Hawaii is used to desalinate water, cool buildings, and supply nutrient-rich water for vegetables and aquaculture, all while generating 210 kilowatts of power. In tropical, coastal areas of the developing world, ocean thermal energy appears to be an especially promising technology, with the potential to reduce reliance on expensive diesel generators.[31]

★   ★   ★   ★

Given the rocky development that solar thermal energy has suffered so far, it is difficult to anticipate the course of future developments. Government bureaucrats appear to have retained their uncanny ability to invest in the wrong technology, and low fossil fuel prices have cooled the interest of utilities. Still, the attractions of a new pollution-free power source, coupled with the niche markets opening up in many developing countries, may be sufficient to spur a takeoff.

One possibility that is being explored in the U.S. Southwest is to convert existing coal-fired power plants into solar thermal power stations that use natural gas for backup—already close to being economical for plants that face legal requirements to cut the amount of sulfur and nitrogen oxides they emit, according to one set of calculations. Another possibility is to link a field of solar troughs or dishes to a gas turbine power station, allowing the solar collectors to substitute for the boiler in a combined-cycle plant, which would increase the overall efficiency.[32]

The sheer abundance of solar energy suggests that it will be the foundation of a sustainable world energy system a century from now. Indeed, if we could harness just one quarter of the solar energy that falls on all the world's paved areas, we could meet all current world energy needs comfortably. Moreover, according to several detailed studies, solar thermal technologies should be able to provide power at 5–7¢ per kilowatt-hour by 2000, which could be broadly competitive with gas-generated electricity. Future costs may be lower still.[33]

Solar thermal energy will have to compete against many other renewable energy technologies, however, including not only wind and biomass energy but also its solar cousin, photovoltaic cells, which are discussed in the next chapter. Still, as a source of baseload and peak electricity in relatively sunny, dry climates, solar thermal power is likely to have few rivals. If necessary, there is no reason that electricity or hydrogen produced at solar power plants in remote areas cannot be conveyed to homes and factories located in distant cities. The eventual role of solar thermal energy will in the end be shaped by the way the broader energy economy evolves.

8

Plugging into the Sun

In Joba Arriba, a scattered farming settlement of about 5,000 people on the Dominican Republic's northern coast, Manolo Hidalgo decided to give his family a different kind of Christmas present in 1991: electric lighting. No longer content with dim and smelly kerosene lamps to light the homework of his three children, Hidalgo knew that his chances of getting electricity in the conventional way—by hounding government officials in the hope they would extend the nearest power line the necessary 15 kilometers—were faint at best. Funding for rural electrification was perennially scarce and government promises unreliable.[1]

Hidalgo came up with a faster, more reliable option. Travelling to the nearby town of Sosua, he visited a small store called Industria Eléctrica Bella Vista, which offered a self-contained solar panel that converts the sun's rays directly into electric current. With help from the store's electrician, Hidalgo mounted the roughly one-meter-square panel on his roof, and wired it to a lead-acid battery in order to store electricity for use in the evening. The day after visiting Sosua, his home had five working electric lights, as well as a radio and television set, all powered by his solar photovoltaic (PV) panel. With the flip of a switch on that December day, Hidalgo and his family joined tens of thousands of rural people who in the past few years have begun getting their power directly from the sun.

It is hard not to be struck by the irony that the basic energy needs of some of the world's poorest people are being met by what is arguably the most elegant and sophisticated energy technology yet developed. Indeed, solar photovoltaic cells, which can be used in everything from handheld calculators to suburban rooftops and large desert power stations, are a unique and potentially revolutionary means of generating power. Solar cell technology is advancing rapidly, and many experts expect the devices to be ubiquitous in the early part of the next century. But the greatest short-term impact of solar photovoltaics will be in the rural Third World, providing power to many of the more than 2 billion people like Manolo Hidalgo who do not yet have it.[2]

Solar photovoltaic cells are semiconductor devices made of silicon—similar to but far less expensive than the chips used in computers—that convert the energy from sunlight into moving electrons, avoiding the mechanical turbines and generators that provide virtually all the world's electricity today. French scientist Edmund Becquerel discovered in 1839 that light falling on certain materials could cause a spark of electricity—

known as the "photoelectric effect"—and that this charge could, under the right conditions, be propagated, forming an electric current. Within 50 years, scientists began to manufacture primitive solar cells out of a rare element known as selenium. The high cost and low efficiency of these devices made them useful in only one serious application—photographic light meters.[3]

The precursors of the modern photovoltaic cells used on Manolo Hidalgo's roof did not emerge until 1954. Early that year, a small team of scientists at the Bell Laboratories in New Jersey began looking for a way to generate electricity for telephone systems in remote areas. As they tried to improve the selenium cells, a separate Bell team discovered that a silicon device they were testing produced electricity when exposed to sunlight. When the work of the two groups was merged, a breakthrough was achieved: a silicon cell that converted 4 percent of all incoming sunlight into electricity—five times as much as the best selenium cell. Within a few months, the Bell team had pushed the cell's efficiency to 6 percent. Excitement about this development was compounded by the fact that silicon is the second most abundant element, constituting 28 percent of the earth's crust. Soon, the Bell scientists began to speculate that silicon solar cells might become a major source of electricity, and their enthusiasm was echoed by the popular media, which touted the imminent arrival of a solar-powered future.[4]

Realizing that potential turned out to be a far greater challenge than anticipated. The initial silicon solar cells cost roughly $600 per watt ($3,000 in 1993 dollars), several thousand times as much as electricity from conventional plants. After building a few prototype solar panels, Bell Labs decided that no commercial applications were within reach, and shelved the new technology. The photovoltaic cell was saved from obscurity by the U.S.-Soviet space race in the sixties. In the rush to find a practical way to power satellites, U.S. space scientists dusted off the solar cell.[5]

During the next decade, a sizable infusion of government funds served to jumpstart the photovoltaic industry, which was led by electronics companies such as RCA, Texas Instruments, and Heliotek. Within 10 years the price of solar cells was cut by a factor of 5 to 10, as cell efficiency rose and the durability of the devices improved. Producing these early PV cells required pure, expensive silicon and wasteful, energy-intensive manufacturing processes. Individual wafers of silicon were sliced from long crystals grown at temperatures above 1,400 degrees Celsius. Although costs were beginning to fall, PVs were still too expensive for any application other than the government-funded space program.[6]

The 1973 Arab oil embargo created a new race—to bring photovoltaics back to earth. Government energy agencies and scores of private companies invested billions of dollars in advancing the state of photovoltaic technology. By 1980, the efficiency of commercial PV modules had risen to more than 10 percent and the price had fallen to roughly $12 a watt ($21 in 1993 dollars). (See Figure 8–1.) Finally, silicon PV cells were cheap enough to fill the niche Bell Labs originally tried to create for them. During the eighties, solar cells were widely deployed at telephone relay stations, microwave transmitters, remote lighthouses, and roadside callboxes— applications where small amounts of power are needed and where conventional power sources are either too expensive or not reliable enough.[7]

The technology continued to advance during the next

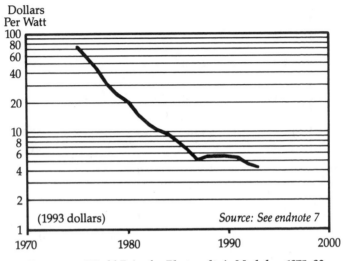

FIGURE 8-1. *World Price for Photovoltaic Modules, 1975–93*

decade, and by 1993 the average wholesale price of photovoltaics had dropped to between $3.50 and $4.75 a watt, or roughly 25–40¢ a kilowatt-hour, thanks both to higher efficiencies and more automated manufacturing processes. As costs fell, sales rose—from 6.5 megawatts in 1980 to 29 megawatts in 1987, to 60 megawatts in 1993. (See Figure 8–2.) The worldwide industry, including ancillary activities such as retail sales and installation, did roughly $1 billion worth of business in 1993.[8]

Although still too expensive to compete head-to-head with conventional generating technologies, photovoltaic cells have found ever-larger niches in the global energy market. The technology's versatility was best demonstrated in the mid-eighties, when Japanese electronics companies came up with a particularly ingenious application, attaching tiny solar cells to small consumer de-

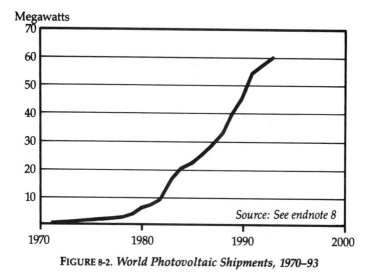

FIGURE 8-2. *World Photovoltaic Shipments, 1970–93*

vices such as handheld pocket calculators and wrist-watches. These require only a trickle of electricity, well within the capability of a small solar cell—even when operating in a dimly lit room. Since the late eighties, the Japanese have sold an average of about 100 million such devices each year, an application that absorbs 4 megawatts of solar cells annually, 6 percent of the global market. The use of solar cells in consumer electronics has levelled off in recent years, as has that in telecommunications, surpassed by more rapidly growing applications ranging from rural water pumps to village homes and rural electricity.[9]

By the early nineties, thousands of villagers in Africa, Asia, and Latin America were doing just what Manolo Hidalgo did, using photovoltaic cells to power lights, televisions, and water pumps, needs that are otherwise met with kerosene lamps, rechargeable batteries, and diesel engines. More than 200,000 homes in Mexico, Indonesia, South Africa, Sri Lanka, and other developing countries have obtained electricity from rooftop-mounted solar systems over the past decade. Most of these efforts have been pioneered by nongovernmental organizations and private businesses, with only limited support by government and aid agencies. Earlier aid-funded efforts in the early eighties to electrify whole villages using photovoltaic systems were expensive and had too little local involvement to be sustained. Many broke down and were abandoned soon after the foreign technicians left.[10]

Perhaps the best example of the new approach to solar electrification is in Hidalgo's Dominican Republic, where more than 2,000 homes have been "solarized" in the past nine years, largely through the efforts of Enersol Associates, a U.S. nonprofit group founded by Richard Hansen, and a Dominican organization called the Asociación para el Desarrollo de Energía Solar. A former Westinghouse engineer who once designed equipment for coal and nuclear power plants, Hansen rediscovered his earlier interest in solar energy during vacations in the Dominican Republic. In early 1984, armed with a small PV panel and a vision of bringing electric light to the campesinos, Hansen set off for the Caribbean. Eschewing the bureaucracy and paperwork of government and foreign aid institutions, he soon sold his single unit—on credit—to a rural family who wanted electric lighting for the small store it operated.[11]

Soon a handful of other families were asking Hansen how to purchase their own systems. A year later, the families banded together to form a revolving credit fund, using $2,000 in seed money from the U.S. Agency for International Development, and purchased five more solar systems. As it has grown, the fund has offered PV loans to more families each year. Participants make a down payment of at least 12 percent, and then pay monthly installments over two to five years. The funds are recycled into new PV loans.[12]

Since 1990, similar organizations have been established in China, Honduras, Indonesia, Sri Lanka, Zimbabwe, and elsewhere. In Sri Lanka, $10 million is enough to electrify 60,000 homes, estimates Neville Williams, founder of the Solar Electric Light Fund, a U.S.-based nonprofit that facilitates PV electrification in Asia and Africa. Still, at early-nineties prices (roughly $500 for a 50-watt system that includes not just a PV panel but also light bulbs, wiring, a battery for storage, and a charge regulator that monitors the battery), only a few people can afford PVs. A survey by Hansen estimated that just 20 percent of the rural households in the Dominican Republic that do not yet have electricity can now afford a system, even with a low-interest loan. Further price cuts will increase the number, as will smaller systems being designed specially for Third World village use. A recently developed solar lantern, for example, uses a 2.6-watt panel to charge a battery that can light one or two fluorescent lamps, an example of a new product that can provide basic energy services even to very poor families.[13]

Still, solar electrification projects start with a large disadvantage, since most developing-country governments heavily subsidize the extension of grid electricity to rural areas, as well as the installation of diesel water pumps. Simply levelling the playing field—reducing the subsidies to conventional power or providing equivalent funding for solar energy—could lead to a boom in solar electrification. Slowly, a growing number of Third

World governments and international aid agencies have begun to respond to this need, mainly by setting up new ways of funding solar power projects.[14]

Some of the recent impetus has come from the Global Environment Facility (GEF), a fund set up in 1990 under the joint management of the World Bank, the United Nations Development Programme, and the United Nations Environment Programme to finance projects that are not quite economical today but that will benefit the global environment by keeping carbon dioxide and other pollutants out of the atmosphere. In Zimbabwe, a $7-million GEF grant approved in 1992 will finance a revolving fund that will be used to electrify 20,000 households in five years; another $55-million World Bank loan and GEF grant will support a program to install 100,000 solar lanterns and undertake other projects in India. The World Bank is also considering solar loans to China, Indonesia, the Philippines, and Sri Lanka. Some of the new grants and loans are to strengthen nascent PV industries in developing countries, which can create economic opportunities and jobs in rural areas. The Zimbabwe grant, for example, is supporting six small PV installation companies and a larger enterprise that imports cells and assembles them into commercial panels.[15]

The use of solar electric systems in rural homes is growing in industrial countries as well, spurred by the popularity of vacation cabins and the cost of reaching them with power lines, which in the United States runs between $13,500 and $33,000 per kilometer for even small local distribution lines. By contrast, a 500-watt PV system—enough to power an efficient home's lights, radio, television, and computer—would cost less than $15,000, including batteries for storage. Norway already has 50,000 PV-powered country homes, and an additional 8,000 are being "solarized" each year. Among the other leaders in PV home installations are Spain, Switzerland, and the United States. All four nations have extensive forests or mountains, and a middle class with the money and leisure time to enjoy them.[16]

Electric utilities are beginning to serve the remote home market as well, in effect redefining their structure to include potential users who are not actually connected to the utility's web of power lines. In the rugged mountains and remote basins of the northwestern United States, the Idaho Power Company is purchasing, installing, and maintaining PVs for homeowners located off the grid. The utility also helps customers select energy-efficient appliances in order to keep the initial costs of the solar system down. Instead of a monthly bill based on electricity use, customers pay a set monthly fee based on the cost of installing and maintaining the system. Southern California Edison and the Western Area Power Administration have similar programs, as do several of Brazil's state-owned utilities, which have found it impossible to meet government electrification goals any other way.[17]

★　★　★　★

Prospects for using solar photovoltaics in ever-wider applications hinge on how rapidly the technology evolves. Prices will need to be cut by a factor of three to five in order for large-scale grid-connected applications to become economical. Most PV experts are confident that such reductions can be achieved by continuing the advances in cell efficiency (see Figure 8–3) and manufacturing processes of the past two decades and by capturing the cost-saving potential of mass production.

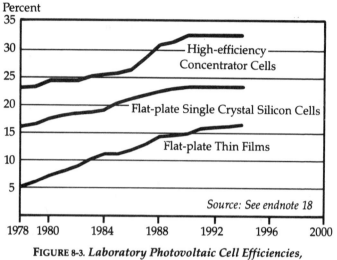

FIGURE 8-3. *Laboratory Photovoltaic Cell Efficiencies, 1978 Through February 1994*

Experts disagree vehemently, however, about which of the many promising PV technologies will make it, and which markets will provide the needed bridge to the future.[18]

As of the mid-nineties, single-crystal silicon is still the dominant PV technology, with 48 percent of the global market. Single-crystal panels are reliable and durable, and manufacturing efficiencies continue to improve, which is projected to cut costs significantly in the late nineties. High costs for materials and processing will prevent this technology from falling to much less than half of current prices, however. A related material, poly-crystalline silicon, which is made up of millions of tiny crystals, had captured 30 percent of the world market by 1993. According to experts, it has somewhat greater potential for cost reduction than do the single-crystal cells.

Several other promising solar electric devices are starting to leave the laboratory and go into commercial production in the mid-nineties, however, with the potential to use less expensive materials, reach higher efficiencies, and greatly lower manufacturing costs.[19]

One approach being pursued by researchers is to develop highly efficient solar cells made of materials such as gallium arsenide, which have already reached efficiencies as high as 33 percent in the laboratory. Such cells would be used in conjunction with lenses and reflective mirrors that focus the sun's rays onto them, greatly reducing the amount of semiconductor material needed. The array is usually mounted on a special dual-axis tracking device that allows it to be continuously pointed toward the sun. Because concentrators are unable to use diffuse sunlight, they are most attractive in areas of high and direct insolation, such as deserts and other arid regions. Some experts believe that these concentrators have the greatest potential for achieving an economical utility-scale solar power plant, though it should be noted that solar thermal power technologies (see Chapter 7) currently have a sizable economic advantage over PVs in this application.[20]

Another promising technology appeared in April 1991, when Texas Instruments (TI) announced plans to produce a "Spheral" cell made of tiny silicon beads that is projected to cut costs to less than $2.50 per watt. The key to TI's cost projections is its use of metallic-grade silicon (which costs around $1 a pound) instead of the purer semiconductor-grade silicon (more than $50 a pound). TI plans to begin commercial production in the mid-nineties. Another recent innovation is a cell invented at the Swiss Federal Institute of Technology that is made of ruthenium and titanium oxide and that mimics the photosynthetic process of plants. Scientists believe that the cost of this "nanocrystalline" PV cell will be one fifth that of today's cells, an estimate that has led two large corporations, Asea Brown Boveri and Sandoz, to collaborate with the Swiss institute.[21]

The new photovoltaic technology that has captured the most scientific attention and the largest share of the "unconventional" PV market is the so-called thin-film solar cell. Less than 1 micrometer thick (compared with 300 micrometers for a crystalline cell), thin films require much less raw material and can be applied directly onto glass or sheets of metal foil, thereby simplifying the manufacturing process and producing a panel that can be easily stored and transported. Scientists at the National Renewable Energy Laboratory in the United States say that by the late nineties, polycrystalline thin-film modules may achieve commercial efficiencies of about 15 percent.[22]

Another thin-film material that has garnered attention in recent years is amorphous (or noncrystalline) silicon. With a composition similar to common sand, amorphous silicon is inexpensive, but with careful processing, it, too, can be used for producing electricity. Moreover, amorphous silicon can be processed at a low temperature, which allows such modules to generate as much energy in their first year as is required to produce them. Although amorphous silicon solar cells were first conceived of by the maverick U.S. inventor Stanford Ovshinsky in the sixties, their initially low efficiency levels discouraged interest. The Japanese found the first commercial application in the eighties, in the consumer electronics devices mentioned earlier.[23]

Researchers encountered another problem with the early amorphous silicon photovoltaics. During their first

1,000 hours or so of exposure to sunlight, the efficiency of such cells fell by as much as half. The decay became worse when researchers tried to build thicker, more-efficient PVs. However, Ovshinsky's Michigan-based company, Energy Conversion Devices, Inc. (ECD), has gotten around the problem by combining two and even three layers of amorphous silicon while minimizing the decay to roughly 15 percent. Researchers at ECD and other companies "dope" each layer with a chemical that allows it to absorb specific wavelengths of sunlight, thereby capturing more of the sun's rays.[24]

After a decade of work, ECD announced the development of a triple-junction PV module in 1994, which after the initial degradation still has an efficiency of 10.2 percent—an achievement thought nearly impossible a decade ago. ECD plans to manufacture these panels in a joint venture with the Japanese company Canon, under the name the United Solar Systems Corporation. According to company literature, the factory will be able to produce 10 megawatts of PV panels annually, and those panels will generate power at a cost of about $1 per watt or 10–12¢ per kilowatt-hour generated—less than a third of the average cost of PV electricity in 1993.[25]

Other companies appear to have similar targets in mind. Advanced Photovoltaic Systems, AstroPower, Golden Photon, and Solar Cells, Inc., are among the U.S. companies building larger factories to produce a variety of new, thin-film solar cells, some using materials such as cadmium telluride and copper indium diselenide. Unlike the labor-intensive batch processes now used by most PV manufacturers, future production lines will incorporate smart automation and assembly-line techniques that will raise yields and lower production costs. The color film industry produces bulk film at roughly $5 per square foot through manufacturing techniques analogous to thin-film PVs. This is equivalent to producing bulk PV cells at 50¢ a watt, according to Jim Caldwell, former president of ARCO Solar and now technical director of the Center for Energy Efficiency and Renewable Technologies in Sacramento, California.[26]

During the past two decades, many large corporations have been drawn to the high-tech lure of a mass market for solar cells, but only a few have shown staying power. Among the early investors were a number of major oil companies that decided in the seventies that if they wanted to be in the energy business for the long haul, they should be ready to make the transition from petroleum to sunshine. Atlantic Richfield, British Petroleum, Exxon, and Mobil moved in, causing widespread alarm about the prospect of the oil industry controlling a new global business in the twenty-first century.

The deep pockets of Big Oil were not matched by its strategic vision, however. After posting consistent losses, most of the oil companies found they were not well equipped for a business dominated by high-tech manufacturing and complex sales strategies, and they sold out. Among the other large U.S. firms that entered the field and then abandoned it are General Electric and Westinghouse, which were looking for long-term replacements for their steam generator markets, and IBM, Polaroid, and RCA.[27]

Even after all the financial carnage, at least 30 companies manufactured PVs in the mid-nineties—more than ever before—including BP Solar in the United Kingdom, AMOCO's Solarex Corporation in the United States, and NAPS in France. The world's leading PV producer in 1993 was the German electronics giant Siemens, which purchased ARCO Solar of California in

1990. Among the other key players in the global PV market are well-known Japanese corporations such as Sanyo, Kyocera, and Sharp, which have invested in a range of photovoltaic technologies. Typically, the major PV companies have manufacturing plants in their home countries and extensive international sales operations, particularly in the fast-growing markets of Africa, Asia, and Latin America.[28]

Many developing countries would like to get into the photovoltaics business as well, lured by the fact that their consumers will make up a large share of the growing global market. While solar cell manufacturing is a complex process requiring advanced technology, assembling panels from purchased cells requires only basic electrical engineering skills. Mexico, Morocco, Sri Lanka, and Zimbabwe already have domestic industries putting together photovoltaic panels. In Kenya, Tanzania, and Zambia, the indigenous PV industry still concentrates on installing imported systems, but with outside support, they too may develop indigenous assembling capabilities.[29]

Joint ventures that gradually increase the amount of "domestic content" seem like the best way for most developing countries to build up indigenous industries. It is likely that many of these countries will one day export assembled panels to the industrial North, as Mexico already does. India, which along with Brazil and China has tried to develop a government-subsidized PV industry from scratch, only produces small quantities of solar panels. In recent years, even the Indian government has begun to take a new tack, encouraging domestic companies to form joint manufacturing ventures with some of the leading PV companies in Europe and the United States.[30]

★   ★   ★   ★

Industry observers expect the PV industry to remain dynamic and competitive in the decade ahead as companies now engaged in research and development enter the commercial market, and as others scale up production, generating a fierce price war. In photovoltaics, as in computers, companies that remain satisfied with the status quo will not only lose market share but may find themselves drummed out of business entirely by more dynamic competitors. Part of the race is technological, but it is also one of selecting which market path will best lead to a profitable business.

Although the remote power market will expand, and will have an enormous impact on people, it is dwarfed by the $800-billion annual global market for grid-connected power. Even if 4 million homes (1 percent of those now lacking electricity) installed a PV system each year, a total of only 200 megawatts—less than four times world production in 1993—would be produced. Meanwhile, the world's electric utilities install 60,000 megawatts of generating capacity each year. If the PV industry were able to garner even 1 percent of this market, annual production of solar cells would rise tenfold. Capturing 10 percent of this market would allow a hundredfold increase in production.[31]

Electric utilities have been experimenting with photovoltaics since the early eighties, when a few large projects were installed. Though not economical, these first systems operated with excellent reliability and low maintenance costs. Early on, utility engineers approached PVs as just another generating option, to be deployed in large, central stations once costs became competitive. Although this approach has deep roots in the culture of electric utilities, it fails to exploit the unique advantages of photovoltaics—advantages that could offset some of their costs.[32]

Some utility engineers now believe that by strategically integrating solar cells at bottlenecks in the transmission and distribution system where additional power needs cannot be met without expensive new equipment—and where fossil-fueled plants are not feasible—PVs can be economically justified today. In addition, research has shown that in many climates, photovoltaics provide the greatest amount of electricity just when it is needed most—on summer afternoons when air conditioners are going full blast. This "peak shaving" capability of solar electricity greatly increases its value compared with a centralized power station.[33]

These benefits were documented by the Pacific Gas & Electric Company (PG&E) when it installed a 500-kilowatt PV plant at its Kerman substation in California's San Joaquin Valley in 1993. The system avoids an expensive upgrade of the substation and transmission lines that otherwise was needed, while improving reliability. PG&E engineers estimate the value of PV power at the Kerman station at two to three times that of centrally generated electricity. They believe that the company could deploy more than 120 megawatts of distributed, grid-connected PVs. Extrapolated to the United States as a whole, some 4,500 megawatts of grid-connected PVs might already be cost-effective—70 times the current annual global market.[34]

Other utilities are going a step further, installing PVs right into buildings themselves. Over the past decade, several dozen grid-connected solar buildings have been built. (See Chapter 11.) The Sacramento Municipal Utility District plans to purchase an average of more than 1 megawatt of distributed photovoltaic systems each year until the end of the decade. It launched its effort with a 1993 procurement of 640 kilowatts, installed on residential rooftops and at an electrical substation. Southern California Edison, an investor in TI's "Spheral" cell, expects to install the new technology throughout the smog-plagued Los Angeles basin as a way to meet peak power needs without burning fuel or building more power lines.[35]

In order to move into larger markets and continue the downward trend in prices, photovoltaics will need both a push and a pull from governments and utilities. Most utilities are unfamiliar with the idea of installing generating systems in customers' buildings, and they are regulated in a way that discourages investment in technologies that are expensive today but may reap large benefits in the future. In response, some governments have adopted programs aimed at stimulating the market for photovoltaics—to encourage utility investment and at the same time allow private manufacturers to scale up production.

Governments have of course been heavily involved in photovoltaics ever since the early days of the U.S. space program. Support was quickly increased in the seventies, but since then funding has ebbed and flowed in response to public mood swings and the ideological bent of politicians. U.S. government support rocketed to $260 million (in 1993 dollars) in 1980, but then fell to just $38 million in 1990. Government R&D funding was steadier in Germany and Japan during the eighties, exceeding U.S. outlays by the end of the decade. Much of the early government support in all these countries was skewed toward large demonstration projects, however, most of which did the technology little good.

Moreover, the wild swings in U.S. government support made it difficult for the PV industry to attract the private capital needed to advance the technology, reduce manufacturing costs, and bring new products to market.[36]

Between 1990 and 1994, U.S. government R&D support nearly doubled, reaching $78 million, while Germany's plateaued and Japan's grew modestly, putting the United States back in the number one position. At the same time, governments have shown better understanding of the needs of the industry. R&D programs have been reoriented so as to work with private efforts, and governments have also begun working to spur commercial development.[37]

Some of the most effective programs are in Europe, where governments began offering tax incentives, low-interest financing, and cash rebates to entice building owners to install solar electric systems in the early nineties. In Germany, the Thousand Roofs program was launched in 1990, and was soon upgraded to 2,500 roofs. Switzerland aims to place at least one PV system in each of the country's 3,029 villages by the end of this decade.[38]

The Netherlands is planning to install 250 megawatts of photovoltaics by 2010, with the expectation that PVs will contribute far more power after that. Austria and Denmark are preparing their own stimulus programs. Japan, meanwhile, has devised a strategy to install some 62,000 building-integrated PV systems, with a combined capacity of 185 megawatts, by the end of the nineties. Government support will take the form of a subsidy that initially will cover half the cost of the system, but then will decline gradually until it reaches zero after seven years. With their peak power demand coming from air conditioning demand on sunny days, Japanese companies are focused on developing solar-assisted air conditioners.[39]

In the United States, federal and state tax credits have been used to spur photovoltaics since the late seventies, a subsidy that was still in place in the mid-nineties. In 1993, a coalition of more than 60 U.S. utilities announced plans, in conjunction with the federal government, to install 50 megawatts of solar cells between 1994 and 2000.[40]

Still, industry analysts believe that none of these programs is yet adequate to the task at hand. Until the photovoltaics market reaches a scale of perhaps 500 megawatts per year—nearly 10 times the current level—it will remain a specialty business incapable of achieving the low costs associated with mature industries.

In order for manufacturers to scale up production, governments will have to work with electric utilities to organize much larger bulk purchases than currently are being considered. Another option is to purchase photovoltaics directly—for use on government buildings or at military facilities. The U.S. government, for example, through the General Services Administration could spur the PV industry by cost-effectively deploying 500 megawatts on government buildings over the next 10 years, estimates Harvey Sachs of the Center for Global Change at the University of Maryland. The key is to use competitive, multiyear contracts that couple expanding purchases with price reductions in each succeeding year.[41]

With strong and consistent support in the decade ahead, solar electric technology may achieve the economic and commercial breakthrough that the industry has long waited for. Indeed, it is not unlikely that manufacturers could bring the cost of solar electricity down to

10¢ a kilowatt-hour by 2000, or even 4¢ by 2020. If so, photovoltaics could become one of the world's largest industries, as well as one of its most ubiquitous energy sources.[42]

9

Plant Power

During the winter of 1994, engineers from the Finnish engineering firm Ahlstrom and the Swedish power utility Sydkraft commissioned a new heating plant in the southern Swedish town of Värnamo. The new facility uses a jet-age technology but is fueled by an energy source first captured by our ancestors when they harnessed fire more than a half-million years ago: wood. The Värnamo plant generates 6 megawatts of power and 9 megawatts of heat for the town's district heating system by gasifying the wood and burning it in a jet engine. In all, more than 80 percent of the energy embodied in the wood will end up heating buildings or powering lights and motors in the town, while emitting no sulfur and only as much carbon dioxide as is absorbed by the new trees planted to replace the ones burned.[1]

The Värnamo plant is symbolic of a new generation of

technologies that are bringing the most ancient energy source rocketing into the twentieth century. Recent advances in combustion engineering, biotechnology, and silviculture are making it economical to turn a variety of plant forms into usable liquid or gaseous fuels, or even into electricity.

Although the theoretical potential of biomass energy is undeniable, questions remain as to just how practical an energy source it will become. Some experts maintain that the world has enough agricultural and forestry resources to make biomass the foundation of the world energy economy of the twenty-first century; a study commissioned by the United Nations for the 1992 Conference on Environment and Development found that if crops were grown specifically for this purpose, the equivalent of 55 percent of today's total world energy use could be met with biomass by 2050. Whether such visions can ever be practically realized depends on the availability of productive land and adequate water and fertilizer—resources that may be in short supply in the decades ahead.[2]

Harnessing biomass energy is in effect a way of exploiting nature's solar collectors—living plants that use photosynthesis to turn the energy of sunlight into carbohydrate molecules. Animals are able to "borrow" some of that energy when they eat vegetable matter, and throughout history, people have relied on the energy in biomass when they burn wood in order to cook food or heat their homes. Since the emergence of fossil fuels as a major energy source late in the nineteenth century, biomass has been largely ignored by energy planners and statisticians alike, in part because much of its use falls outside the commercial markets that governments keep tabs on.

Although U.N. data indicate that biomass supplies

just 5 percent of the world's energy, a more complete accounting by independent experts found that this source actually provided 13 percent of the world's energy in 1992. In developing countries, biomass currently supplies an estimated 36 percent of the energy used, and it provides virtually all the fuel used by an estimated 2.5 billion people who mainly live in rural areas—45 percent of the world's population. Even in some industrial countries, such as Denmark and Finland, biomass accounts for roughly one tenth of the energy used. (See Table 9–1.) Of course, biomass also provides people with construction materials, food, fodder for animals, and paper—all valuable products that often compete with energy as a claim on wood and other biomass material.[3]

Although biomass is a renewable source, much of it is

TABLE 9-1. *Biomass Energy Use in Selected Countries, 1987*

Country	Biomass Use	Share of Total Energy Consumption
	(petajoules)	(percent)
United Kingdom	46	< 1
United States	3,482	4
Denmark	84	9
Thailand	206	20
Brazil	1,604	25
China	9,287	28
Costa Rica	31	32
Zimbabwe	143	40
India	8,543	56
Indonesia	2,655	65
Tanzania	925	97

SOURCE: See endnote 3.

currently used in ways that are neither renewable nor sustainable. In many parts of the world, firewood is in increasingly short supply as growing populations convert forests to agricultural lands and the remaining trees are burned as fuel. The resulting fuel shortages have forced women and children to spend much of their time collecting wood, while in many areas, crop residues and animal dung—which otherwise are valuable fertilizers— are burned in cookstoves. Biomass energy use in industrial countries is not always sustainable either. As a result of poor agricultural practices, soils in the U.S. Corn Belt, which supplied some 4 billion liters of ethanol fuel in 1993, are being eroded 18 times faster than they are being formed. If the contribution of biomass to the world energy economy is to grow, technological innovations will be needed, so that biomass can be converted to usable energy in ways that are more efficient, less polluting, and at least as economical as today's practices.[4]

Gains in conversion efficiency are already being achieved, starting with cookstoves. Traditional cooking methods often involve no more than a pot supported by three stones, and generally capture less than 10 percent of the wood's energy. At the same time, they emit carbon monoxide, particulates, and a variety of cancer-causing chemicals into poorly ventilated kitchens, poisoning the air for some 400–700 million people, and affecting women and children disproportionately. Improved stoves can effectively harness for cooking more than 40 percent of the energy in wood or charcoal, while reducing dangerous emissions. Some half-million efficient cookstoves have already been sold in Kenya and 130 million in China. Likewise, technologies to transform biomass to electricity currently operate at 20-percent efficiency or lower, yet a scaled-up version of the

Värnamo combined-cycle gas turbine (similar to the natural-gas-based generators discussed in Chapter 5) can convert more than 45 percent of the wood's energy to usable electricity. Further improvements are likely in the future. The net effect, for both wood stoves and power plants, is to provide more than twice the amount of energy with the same amount of fuel.[5]

Although conversion efficiency is important, the greater challenge in relying heavily on biomass energy is providing sufficient quantities of fuel in a way that does not jeopardize the underlying resource base of soil and water. Two main strategies beckon: reliance on residues from crop and forestry production, and the development of new, sustainably grown energy crops. The first of these has the greatest near-term potential. Roughly three quarters of the standing volume of trees harvested from forests in developing countries, for example, ends up as unsalable residues. Recovering all such "waste" materials from agriculture and forestry could one day supply enough energy to meet 7.5 percent of current world energy needs, estimates David Hall, a biologist at the University of London's King's College.[6]

Today, the U.S. paper industry meets roughly half its energy needs using sawdust, scrap wood, and pulping waste to fuel its boilers for heat and electricity generation. U.S. grid-connected, biomass-fueled electricity capacity—all of it based on residue products—rose from 200 megawatts in 1979 to some 6,000 megawatts by 1993. The catalyst behind the increase was the 1978 passage of the Public Utility Regulatory Policies Act, which required electric utilities to buy power produced by independent companies at fair prices. (See Chapter 12.) More recently, U.S. utilities have shown interest in burning wood, which is very low in sulfur, along with coal in order to meet sulfur emission standards, spurring R&D by government and industry. Surveys indicate that an additional 8,000 megawatts of biomass energy potential can be found in the U.S. paper and wood products industries alone.[7]

Denmark, a country with extensive cropland but few large forests, has made straw—its largest agricultural waste product—into an important energy source. Existing straw surpluses can meet more than 7 percent of the country's energy needs. Following the decentralized approach it blazed in developing wind energy, the Danish government has helped spur the construction of 12,000 small-scale straw burners that provide heat for on-farm use. And since 1980, more than 60 district heating systems have been modified to rely on straw for an average of 90 percent of their fuel.[8]

With the help of higher taxes on fossil fuels, Denmark plans to increase straw combustion from 800,000 tons in 1991 to 1.2 million tons in 2000. Ash from some of the plants is being used by farmers to fertilize their fields, thus reducing but not eliminating nutrient loss. As with wind turbines, Danes hope to make biomass technology into an important export: in 1993, a straw-fired unit was sold to the district heating system in the former East Germany, replacing a highly polluting, coal-fired unit.[9]

Residues from sugarcane—produced in more than 100 countries—could add to the energy supply in many areas. Some bagasse, the residue that remains from extracting sugar from cane, is already burned in boilers to produce steam to fuel the extraction process. A few sugar mills also produce electricity with the steam, but usually use an antiquated technology, with a typical sugar mill boiler producing enough steam to fuel the plant's operation and generate tiny amounts of electricity. Modern steam turbines operating at higher pressures, already in use in Hawaii, Mauritius, and a few other areas, can maintain steam production while raising electricity output eightfold.[10]

Even larger increases are expected by converting the solid biomass into a gas (a process that involves rapidly heating it in an oxygen-depleted atmosphere) and burning it in a gas turbine. Biomass-fired gas turbines—like the prototype at Värnamo—would raise electricity output by more than thirtyfold over today's standard boiler (though steam production would fall slightly). If sugar mills burned all their residues using advanced gas turbines today, they would produce an amount of electricity equivalent to one third the total current output of utilities in developing countries.[11]

An international effort, supported by the Global Environment Facility and the Brazilian government, is sponsoring a design competition for a 25–30 megawatt biomass-fueled gas turbine in Brazil's northeast state of Bahia. The project, seen as the next step in scaling up the technology demonstrated at the Värnamo plant, is expected to lead to a facility that can cut the cost of biomass electricity from roughly 8¢ per kilowatt-hour to 5¢ (using biomass fuel costing $40 a ton), making biomass-fired electricity competitive with conventional coal-fired power plants.[12]

Even with existing, inefficient systems, sugar mills have started selling electricity produced from wastes to local power companies in Costa Rica, Cuba, Fiji, Guatemala, Mauritius, Thailand, the United States, and Zimbabwe. The key is to require electric utilities to pay fair prices for surplus power, still a rarity in many countries. Gas turbines also can be fired with other agricultural residues, with forestry wastes currently burned at paper and pulp factories, and even with crops grown specifically for energy conversion, opening up a range of potential applications in virtually every country.[13]

Other residues, such as dung from livestock and humans, can be gasified in anaerobic digestors, with the remaining nutrient-laden material returned to the land as a rich fertilizer. Digestors are usually either brick or metal tanks inside which fermenting organic materials give off flammable gases such as methane, a process quite different from the heat-driven gasification step used for fueling gas turbines. Methane created by microbes that decompose landfilled garbage can also be collected and burned. China, Denmark, India, and the Netherlands, among other countries, are using large and small anaerobic digestors to convert dung to energy, while the United Kingdom has 51 landfill plants, which capture 8 percent of the amount of the methane generated—and released into the atmosphere—each year. And in western Germany, a quarter of the landfills are equipped to convert the gas to heat and power.[14]

Burning methane from landfills has an additional advantage: the digestors consume a gas with 11 times the greenhouse strength (per molecule) of carbon dioxide. A study by energy analysts at the Pacific Institute for Studies in Development, Environment, and Security found that when a landfill power plant produces a kilowatt-hour, it burns enough methane to offset the carbon released when 10 kilowatt-hours are generated by a coal-fired plant.[15]

Burning is not always the best way to deal with biomass refuse, however. Some of the materials coveted by energy developers, such as paper, have a higher energy value if they are recycled or reused rather than burned.

Recycling paper saves two to four times as much energy as can be produced by burning it. Waste incinerators in the United States are already experiencing fuel shortages due to successful recycling and waste-minimization efforts. Moreover, continued efforts to improve the efficiency with which biomass products are processed will reduce the amount available for energy conversion. Southeast Asian mills convert as little as 40 percent of raw wood into final products at the moment, while a typical industrial-country rate is 50 percent and in Japan, mills go as high as 70 percent.[16]

Beyond residues, the next step in expanding reliance on biomass energy is to produce crops specifically destined for this purpose. So far, experience is limited, as inexpensive fossil fuels have discouraged farmers from growing energy crops. In a few cases, heavy government subsidies have spurred limited reliance on biofuels as a substitute for oil.

In the United States, nearly a tenth of the motor fuel used in the mid-nineties is a 10-percent blend of corn-derived ethanol and gasoline, a product that has been made economically attractive through exemptions from federal and state fuel taxes. In Brazil, pure ethanol—the same substance found in alcoholic beverages—is distilled from sugarcane and used to power about a third of the country's cars. Together with ethanol that is blended with gasoline, alcohol now provides about half the fuel for Brazil's automobiles. Costs have fallen by roughly half since the late seventies, though the least expensive cane ethanol still costs more than gasoline refined from a $30 barrel of oil—roughly double the world price in early 1994. A few other countries, most notably Zimbabwe, have smaller sugarcane-to-ethanol programs.[17]

Although these technologies will continue to improve,

there is little chance that either corn- or sugarcane-derived alcohol will be economically justifiable in the near future. But the main limit to reliance on conventional biofuels is the availability of cropland. Globally, there is simply not enough grain to fuel the world. If the entire world corn crop were converted to ethanol, for example, it would meet only 13 percent of current global gasoline demand; the world sugar crop could meet only another 7 percent. Moreover, most of today's crops depend heavily on fossil fuels for fertilizer, pesticides, and the energy that goes into plowing, irrigation, harvesting, and processing. Using those crops as an energy source would yield little if any net reduction in fossil fuel dependence or carbon emissions. In other words, long-term sustainable use of biomass fuels might require a virtual reinvention of agriculture.[18]

One modest step in this direction is to develop new energy crops that contain lower-grade and less valuable plant materials, such as cellulose and other complex organic matter. With additional research, scientists believe it will be possible to convert these into carbohydrates and then into alcohol. Wood, for example, contains cellulose, hemicellulose, and lignin, each of which has chemical bonds that resist breaking. The National Renewable Energy Laboratory, however, has developed an enzymatic process that can turn them into ethanol. Already, this has lowered the cost of enzymatic ethanol from $7.10 a gallon of gasoline equivalent in 1980 to $1.30–1.70 in 1994 (in 1993 dollars). Researchers believe the cost can be halved in the near future, bringing ethanol closer to the current wholesale price of gasoline of 45¢ a gallon. Similarly, they continue work on lowering the cost of producing methanol—or wood alcohol—from biomass.[19]

Among the crops that would be more efficient than

traditional foodstuffs in harnessing biomass energy are
perennial grasses, such as switch grass and elephant
grass, and fast-growing trees, such as poplars and wil-
lows. Grasses could be grown and harvested every 6–12
months using methods similar to those used with alfalfa
and other forage crops, while trees would be mechani-
cally harvested every three to eight years. Many of the
tree species under investigation will grow back from existing roots, avoiding replanting, though nitrogen fertilizers (which are currently produced from fossil fuels)
might still have to be applied annually.

Farmers could also intersperse energy crops among
more traditional crops. Upon harvesting, which along
with handling would account for as much as half the
production costs of energy crops, the material would be
trucked to a relatively decentralized conversion facility
to avoid unnecessary transportation of bulky biomass,
where it would be converted either to a liquid or gaseous
fuel or burned directly to generate electricity. One system even envisions burning whole logs, reducing the
processing costs.[20]

According to one set of theoretical estimates, such
techniques might allow biomass to provide 38 percent of
the world's liquid and gaseous fuels and 18 percent of its
electricity by 2050, a figure equal to more than half of
the world's current primary energy use. More daunting
is the hundreds of million of hectares of land that would
be needed. In industrial countries, the 32 million hectares of land currently withheld from food production
(much of it highly erodible) could produce just 5.5
exajoules of energy, equal to 3 percent of industrial-
country energy use in 1991. So although land availability might not pose an immediate limit to bioenergy, it
could become one if biomass energy development were
ever pursued seriously.[21]

Intensive cultivation of energy crops presents other environmental hazards, many of which are already plaguing conventional agriculture. These include heavy dependence on synthetic fertilizers and pesticides, which contaminate surface and underground water, and excessive soil erosion. Also, expanding agriculture can threaten other species when intensively managed monocultures displace natural ecosystems. In a word, biomass energy is only as sustainable as the methods used to produce the feedstock. In the United States, these issues are considered serious enough that a group of electric utility managers, government officials, and environmentalists have formed the National Biofuels Roundtable to come up with principles to guide the development of environmentally acceptable biomass energy systems.[22]

Many experts believe that modern biomass energy systems could help reduce many of the problems posed by agricultural monocultures, and possibly even enhance local ecosystems. To do so, energy crops will have to be perennial, which would limit soil erosion, and be planted so as to buffer sensitive areas such as waterways and marshes from intensely farmed cropland. And by providing an additional revenue stream for farmers, energy crops would stimulate declining rural economies.[23]

Still, it is uncertain just how much land will be available for energy crops, and how widespread energy farming might become. In many developing countries, where high-quality cropland already is in short supply, hundreds of millions of hectares of degraded lands would need to be planted with fast-growing trees and other energy crops. Under one proposed scheme, countries in Africa and South America would become exporters of liquefied biomass on a scale similar to today's Middle Eastern oil kingdoms. But the large commercial plantations needed to achieve this vision would radically alter rural economies and land use patterns. Moreover, such schemes would put severe pressure on many sensitive lands, a pressure that might become unbearable if climate begins to change.[24]

*   *   *   *

Past efforts to entice Third World villagers to plant more trees have often failed, particularly when the undertaking was packaged as an energy project. Villagers are frequently hesitant to plant trees for energy production—even if they are suffering from fuelwood shortages. This stems in part from conflicts about land, which poor villagers often do not own or control. Another problem is that women, who are usually responsible for obtaining cooking fuel, generally favor tree planting. But they are sometimes thwarted by their husbands, who prefer crops that can be readily marketed.[25]

Biomass energy production will only be sustainable if it is developed in ways that meet the needs and concerns of local people. To do so, it will need to be integrated with a broader sustainable agriculture strategy that produces food and raw materials as well as energy. For example, agroforestry techniques, which combine food and wood production, offer a way to boost yields of several important products at once. Research conducted at experimental stations in Kenya and Nigeria in the mid-eighties found that mixing corn and leucaena trees could increase corn production substantially, while yielding at least 5 tons of wood per hectare annually. China has more than a half-million hectares of mixed crops and trees in the North China Plain, with wood harvests from agroforestry netting roughly half those achieved in Africa, and grain output up by 22–30 percent.[26]

Even though agroforestry can supply additional fuel

and other biomass-based products, the poorest of the poor are landless and will continue to face problems meeting their cooking energy needs. Although using biomass to collect solar energy is currently less expensive than using solar energy directly, biomass energy production could result in higher land prices, which could lead to higher food prices and make it harder for poor people to gain access to land or buy food. Indeed, it is difficult to disentangle rural energy issues from the broader issues of landownership and equity.[27]

The availability of sufficient high-quality land is probably the biggest cloud hanging over the plans of biomass energy advocates. The overall efficiency of harnessing solar electricity through biomass is less than 1 percent, compared with 10–15 percent efficiency for solar cells and 15–25 percent for solar thermal power plants. Producing ethanol from sugarcane in Brazil today is roughly eight times more land-intensive than using photovoltaics to produce an equivalent amount of electricity. Biomass energy does, however, have the advantage of being in a form that is relatively easy to store—a trait not shared by most other renewable energy resources.[28]

Still, in a world with limited land capable of producing food, it is unclear how much can be devoted to energy crops. Per capita grain production peaked in 1984, following earlier peaks in cropland per person and irrigated cropland per person. Crop yields are no longer increasing as rapidly as they once were, and a growing number of countries are expected to develop food deficits in the next decade. With the United Nations projecting that world population will exceed 7 billion by 2025, the world's farmers may have difficulty keeping up with burgeoning food demand. Major reliance on biomass energy might well complicate their efforts.

Moreover, any major increases in grain and land prices, which temporary food scarcity might cause, could easily choke off efforts to develop biomass energy.[29]

Biomass energy development could take one of two paths in the future: an intensive, environmentally destructive path or one that could enhance local ecosystems. In the latter, efficiency of production and short-term yields might be partially sacrificed to ensure long-term productivity, safeguard biological diversity, and protect supplies of fresh water. This course would be more sustainable and also seems more likely. It is inconsistent, however, with the massive use of biomass energy envisioned in some recent studies.[30]

Also open for question is exactly how biomass energy might be used in the future. One thing is clear: actually carrying solid fuels to individual urban homes and factories is a practice that is unlikely to be revived, so biomass will first have to be converted to a liquid or gaseous fuel, or to electricity. The latter currently has greater economic appeal: electricity priced at 5¢ a kilowatt-hour is equivalent to oil at roughly $85 a barrel. Later, biomass feedstocks could also be converted to gaseous hydrogen at a higher overall conversion efficiency than is feasible by converting to liquid fuels such as methanol or ethanol. In this way, biomass-derived hydrogen would complement hydrogen produced from solar and wind energy. U.S. scientists have proposed producing hydrogen from switch grass grown on marginal cropland, with plans to use it in fuel cells that provide farms with electricity and heat.[31]

Yet energy derived from biomass will continue to have less value than many other uses of biomass, such as fodder, construction materials, and feedstock for the paper industry. And as fossil fuels are phased out, new uses of biomass—as lubricants, plastics, and chemical feedstocks—may become critically important. This may leave energy near the bottom of the biomass "food-chain"; biomass materials may be so valuable for their other uses that their growth as an energy source will be surpassed by such newcomers as wind and solar energy.[32]

*   *   *   *

Chapters 6 through 9 have discussed four renewable energy sources—wind, solar thermal, photovoltaics, and biomass—that are leading contenders to supply large amounts of energy in the twenty-first century. But even this list does not exhaust the list of potential renewable energy sources. Although some—such as wave and tidal power—are unlikely to be widely used in the foreseeable future, two others cannot be ignored: hydropower and geothermal energy. The first is already an important energy source, but due to environmental limits, it is unlikely to grow rapidly; the second is likely to expand quickly but unlikely to reach the scale of wind or solar power, mainly because it is concentrated in limited areas of the world.

Hydropower actually predates fossil fuels as an economically important energy source. As of 1991, the world had some 643,000 megawatts of hydropower capacity in operation, supplying one fifth of the world's electricity and 6 percent of its primary energy. Hydro generation has doubled since the early seventies, with most of the growth coming in developing countries, which now have 37 percent of the total installed capacity. The United States and Canada have the most hydropower—at 13 percent of the world total each—while other leaders include Brazil, China, and Russia. Tiny

Norway gets 95 percent of its electricity from falling water.[33]

This traditional energy source draws on unique combinations of topography and heavy precipitation; often the best sites either are in mountainous wilderness areas or threaten heavily populated areas. As many of the best sites have been used, and as local people fight to preserve the remaining ones, large hydro projects have virtually ground to a halt in many countries. Even in developing nations, China has been rocked by extensive turmoil over plans to develop the huge Three Gorges reservoir on the Chang Jiang (Yangtze) River, while India has faced similar turmoil over the Narmada River dam. One response to such conflicts is to turn to the development of small hydro projects that do not require the flooding of large tracts of land, an approach that has been particularly successful in China. Still, world hydro development appears likely to slow in the decades ahead, perhaps providing 35 exajoules in 2025, up 50 percent from 1990.[34]

Compared with hydropower, geothermal energy is a relative newcomer to the energy scene. It is also the only renewable energy source that does not rely on direct or recently converted solar energy. Instead, geothermal energy is derived from the earth's heated core—the energy source behind earthquakes and volcanoes. First harnessed to generate electricity in Italy in 1904, geothermal energy is now used in more than two dozen countries to provide heat and generate power. In some areas, like the Geysers in California, a well-placed borehole can bring large amounts of hot steam to the surface, which can then be used to power a turbine and generator—just as in a thermal power plant.[35]

Worldwide geothermal generating capacity was estimated at 7,000 megawatts in 1993, providing 28 percent of the power in Nicaragua, 26 percent in the Philippines, and 9 percent in Kenya in 1991. By the turn of the century, the world total is expected to grow to 15,000 megawatts, with some 40 countries likely to play a role. The overall resource, though, has barely been tapped. Japan, which currently has 270 megawatts of geothermal capacity, has an estimated potential of more than 69,000 megawatts, nearly double the country's current nuclear electric capacity. Other countries, such as Djibouti and Saint Lucia, could reasonably produce all of their electricity from geothermal resources that have already been identified.[36]

In addition to using geothermal energy to generate electricity, many countries use it directly in factories or to heat water or buildings. In 1990, global direct use of geothermal heat reached the equivalent of 11,730 megawatts. Almost a quarter of that was in Japan. China had 2,140 megawatts, up from only 390 in 1985, and Hungary had 1,280 megawatts.[37]

Although geothermal reserves can be depleted if managed incorrectly (and in some cases have been), worldwide resources are sufficiently large for this energy resource to be treated as renewable. In the United States, for example, the Department of Energy estimates that hydrothermal reservoirs—hot water or steam trapped in rock fracture much the way oil and gas are—could in theory provide 2,400 quads of energy, 30 times current annual U.S. energy use, indefinitely. "Hot dry rock" is even more abundant, though harder to tap.[38]

The key barrier to geothermal energy's gaining a big piece of the energy market is likely to be economics. Costs are unlikely to fall as dramatically as the cost of manufactured technologies such as wind and solar generators. Geothermal will therefore probably assume an importent but modest role in the twenty-first century—rising to roughly 5 exajoules in 2025 and perhaps 10 exajoules in 2050.[39]

III

Energy in Society

10

Reinventing Transportation

When the California Air Resources Board met in September 1990, its members found themselves in a difficult position. Some parts of the state had been in violation of national air quality standards for more than a decade, and by 1990, citizens were demanding progress. The federal government threatened to cut off funding for the state's transportation programs if it did not come into compliance. The Air Board's members knew that automobiles were the dominant source of California's smog, and though cars had become steadily cleaner over the years, the tremendous growth in their numbers overwhelmed much of the expected improvement in air quality. To provide the state's citizens with cleaner air, the members of the Air Board felt they had no choice but to approve the world's toughest auto emissions standards.[1]

Although the plan as a whole generated considerable attention for its stringency, one small element reverberated strongly through the world's major automakers the next day: 2 percent of the cars sold in the state in 1998 had to have no emissions at all, with the share rising to 10 percent by 2003. The California zero-emission vehicle (ZEV) standard, as it came to be known, was greeted by a combination of derision and fear on the part of major automakers, who saw it as arbitrary and extreme, requiring that they build a completely new kind of car— but doing nothing to ensure that consumers would buy them.[2]

U.S. auto companies fought hard to kill the ZEV standard throughout the early nineties, but regulators and environmental groups countered that the only way to find out whether an economical zero-emission vehicle could be brought to market was to require it, pointing out that in the past the auto industry had also claimed that catalytic converters and airbags—now standard equipment—would be impossibly expensive. By 1994, the ZEV standard had not only been upheld in California, but had spread to several northeastern states, with lobbying help from the electric utility industry.[3]

To the consternation of skeptics, the years since the ZEV standard was proposed have turned out to be among the most prolific for the auto industry since Henry Ford rolled out the first Model T in 1908. At major auto shows in Tokyo, Paris, and Los Angeles, automotive experts were drawn to a startling conclusion: after dominating personal transportation for more than eight decades, the gasoline-fueled, internal combustion engine–powered automobile may finally be on the way out. Experimental low-emission cars made of advanced composite materials and run on electric drive trains and electronic controls are now being developed by more than a dozen companies. The new designs may be three to five times as efficient as today's, and reduce emissions of regulated pollutants by at least 95 percent and those of carbon dioxide by 75–90 percent.[4]

Although they are not yet ready to admit it publicly (and are still vigorously fighting the ZEV standard in U.S. courts and before state legislatures), the largest automotive companies appear to be coming around to the idea that by soon after the turn of the century, a completely new kind of vehicle will begin to appear in their showrooms—with every likelihood that it will be quieter, more responsive, and cleaner than today's models.

Even with all this promise, important questions remain about how big a dent in the world's transportation problems can be made by a revolution in the design of automobiles, trucks, and buses. The number of motor vehicles on the world's roads has doubled since 1973 and is approaching 600 million in the mid-nineties. (See Figure 10–1.) Most projections show rapid increases in the decades ahead—particularly in Asia. Unless the redesign of the automobile is joined by efforts to spur public transportation and shape communities so that people are less dependent on their cars, the gains from the new technologies may be as fleeting as past efforts were.[5]

* * * *

Technology and public policy often intersect in unpredictable and synergistic ways, and so it was with the zero-emission vehicle. Rapid development of the new technologies would not have occurred without a push from public policy, but the policy itself would have been stillborn without developments that convinced the Cali-

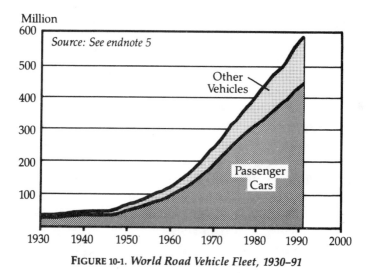

FIGURE 10-1. *World Road Vehicle Fleet, 1930–91*

fornia Air Resources Board that a viable battery-driven car might not be as far off as automakers claimed.[6]

The story began on the morning of June 25, 1985, when 58 odd-looking racing cars pulled out of Romanshorn, Switzerland, and headed toward Lake Geneva, 368 kilometers away. Twenty-seven cars went the distance in the first Tour de Sol—a unique competition in which the only energy source permitted was the sun, and not a gram of air pollution left the cars. Run in normal traffic at a pedestrian average speed of 38 kilometers per hour, the Tour de Sol was intended to demonstrate the potential for solar energy, but its more lasting impact turned out to be on the car itself. The need to rely on the mere 500 watts of solar energy that can be harnessed from a car's roof spurred engineers to press the limits of auto efficiency.[7]

In 10 subsequent runnings of the Tour de Sol, the

cars have gone from Rube Goldberg contraptions to sleek racers with highly efficient electric motors and the latest power electronics, synthetic materials, and aerodynamic designs. The Swiss race and others like it in Australia and the United States now attract top engineering students and auto design teams, stimulating intense competition and rapid technological advances, reminiscent of the turn-of-the-century races that spurred the early development of the automobile. The top solar racing cars have maximum speeds above 150 kilometers (90 miles) per hour, and can go 350 kilometers on a cloudy day before recharging.[8]

These high-tech racers are expensive and, given the vagaries of sunshine, may never be more than a novelty. But the new technology captured the attention of automotive engineers, and soon solar car components began to appear in prototype commercial cars. One of the most important spinoffs occurred when General Motors (GM) entered the World Solar Challenge, in November 1987. GM never does anything on a small scale, and so it brought to this 3,000-kilometer cross-Australia competition its new Sunraycer, a tiny, sleek, $8-million automobile in which the driver lies flat on his back.[9]

Designed with the help of Paul MacCready, a legendary engineer who built the world's first successful human- and solar-powered airplanes, and who now heads the Aerovironment Company in California, the Sunraycer used the latest in highly efficient solar cells and electric motors, and was a full generation ahead of any competitors. As a result, GM beat its nearest rival by 1,000 kilometers. More significantly, MacCready was soon commissioned by GM to work on ways to bring some of the new technology to the commercial market.[10]

Among the experimental vehicles that emerged from

this program was an electric car called the Impact, unveiled to great acclaim at the Los Angeles auto show in early 1990. Roger Smith, then chairman of GM, proudly touted the Impact but also voiced concern that regulators might take it as the go-ahead to require the production of electric cars. Smith did not have to wait long for his nightmare to come true. The California Air Resources Board was already studying the potential of electric cars to improve the abysmal air quality of the Los Angeles basin; in September it voted to adopt the ZEV requirement, effectively cracking a dam that had held back the evolution of automotive technology for decades. Indeed, the most revolutionary aspect of the ZEV mandate is that it forced auto designers to think beyond the incremental improvements in technology they had long pursued.[11]

Although electric cars and a variety of other vehicles were popular at the turn of the century (New York City had a fleet of electric cabs in 1898), they were pushed aside by improvements in the internal combustion engine and the falling price of the gasoline used to run it. For decades, automakers made those engines ever more powerful and efficient, and, when required, they reduced the amount of pollution emitted. Among recent achievements have been a near-doubling of the fuel economy of new U.S. cars between 1974 and 1985 and the addition of electronic engine controls and tailpipe converters to lower hydrocarbon and carbon monoxide output.[12]

When oil prices soared in the seventies, most of the effort to find a replacement focused on the obvious alternatives: other liquid fuels that could be consumed in only slightly modified internal combustion engines. The problem with ethanol and methanol, the two alcohol-based fuels that received the most attention, is that producing them from fossil fuels such as coal or natural gas is relatively inefficient and polluting. Producing alcohol fuels from biological materials such as corn, sugarcane, or wood, on the other hand, is expensive today, and could strain available land and water resources if they ever became a major fuel source. (See Chapter 9.) Methanol, which has been boosted by some politicians, has the additional disadvantage of being highly toxic. Neither is a viable long-term alternative to petroleum.[13]

Rather than focusing on ways to perfect the internal combustion engine or slake its thirst, the most exciting developments since 1990 have leapfrogged over today's automotive technology. Rapid advances in electrical engineering and electronics and the promise of a new generation of improved batteries have convinced a growing number of engineers that a commercially acceptable electric or electric-hybrid car may be within reach. Coupled with advances in lightweight materials, aerodynamic design, and low resistance tires—all of which contribute to greater energy efficiency—these new cars begin to make the current generation of vehicles look downright primitive.

Electric motors are by their nature highly efficient (and becoming more so), and cars with electric drivetrains can easily be equipped with regenerative breaking systems in which the energy of the car's momentum is used to run the motor in reverse, generating electricity that is stored in a battery for later use. In fact, most industrial plants had elaborate mechanical drive systems to run most equipment at the turn of the twentieth century, but these were quickly phased out as inexpensive electric motors were developed. The more compact nature of automobiles helped delay the shift to electricity, but by the nineties, a new generation of motors and electronic controls has made a similar transition economically attractive.[14]

The other key technology for a new generation of vehicles is the lightweight composite materials that were a key ingredient of the solar cars just mentioned. Reducing the weight lowers the energy required to make a vehicle accelerate or climb hills, and increases the range provided by a given battery. As a result, a growing number of companies, including the Swiss chemical giant Ciga Geigy, are adapting lightweight composite plastics—silicon or carbon fibers that are impregnated with polymer resin for use in cars. These materials—already used in bulletproof vests and aircraft—are stiffer and stronger than steel but just one quarter as dense.[15]

Car bodies made of composites emerge fully formed from a mold, much the way recreational boats are made, and can be turned into almost any shape, making it easier to produce curved, aerodynamic designs. Although composites are far more expensive per kilogram than steel is, once production is scaled up, costs will fall. Moreover, composites do not require the complex, energy-intensive bending and stamping that steel parts do, and allow a drastic reduction in the number of car parts, which will tend to cut assembly costs.

Although composite materials involve a number of toxic substances that must be handled carefully during production, they also avoid the need for painting—the single most polluting aspect of automobile production—since colored dye is added directly to the polymer before it is hardened. Recycling these materials is more difficult than for steel, but engineers are already working on recycling strategies, motivated by the fact that some governments plan to require that automakers take back and fully recycle their vehicles.[16]

Conventional wisdom suggests that lighter cars provide less protection in collisions, but design—not weight—is the main determinant of automobile safety. In Switzerland, several very small composite cars have passed standard crash tests, with no serious "injury" to the dummies inside. They are equipped with seatbelts and airbags, and their synthetic bodies are better than steel in absorbing or deflecting the shock of impact. Still, further R&D is needed to ensure adequate safety in the new vehicles.[17]

Many of the advantages of lightweight electric vehicles can be previewed in the GM Impact. With a composite body and stiff aluminum frame, the Impact incorporates an advanced electric motor, electronic controls, regenerative braking, and aerodynamic steamlining that gives it the wind resistance of an F16 figher plane. The energy cost of running an Impact is only a quarter that of a gasoline-powered car, but the driving range is limited to less than 180 kilometers by its 500 kilograms (1,100 pounds) of lead-acid batteries—which hold the energy equivalent of just two liters (a half-gallon) of gasoline.[18]

Some 30 Impacts had been built by early 1994 and were being loaned to consumers for test driving. Auto reporters who have driven it gave the car rave reviews for its quiet ride, maneuverability, and rapid acceleration; GM test-drivers have driven the car at 290 kilometers per hour—a record for electric cars. Even counting the emissions from the power plants used to charge it, the Impact produces one third as much pollution as California allows under its ultra-low emission standard that will be phased in beginning in 1998. Compared with a standard car, the Impact has 30–50 percent fewer body parts, costs one quarter as much for tooling, and requires half as many assembly stations to build.[19]

Although the Impact is a technically impressive car,

many auto industry analysts (and GM executives) are skeptical of the marketability of an expensive car with such limited range. Improved batteries may eventually reduce this problem, but that will take time. Another option that has gained fans in recent years is the so-called hybrid-electric. Since the late eighties, Paul Mac-Cready, Amory Lovins, and others have pointed to the logic of a car with an electric drive that stores the bulk of its energy as a chemical fuel such as gasoline or natural gas—which have energy densities per kilogram that are at least 30 times that of any of the proposed batteries—and that provides a correspondingly greater range. Such vehicles produce most of their own electricity, but carry only enough batteries to provide the extra power needed for acceleration, climbing hills, or, if necessary, operating emission-free inside city limits.[20]

A hybrid-electric car would retain most of the environmental advantages of a battery-powered model, and without the huge quantity of batteries that weigh down the Impact. Whereas today's automobiles turn only 15–20 percent of the energy in the fuel into useful mechanical power, an engine optimized for operation at a single speed could achieve an efficiency of 30–35 percent. And since a hybrid's engine requires only enough power to operate the car at cruising speed, it can be smaller and lighter. (When not needed, the engine would automatically turn off, further reducing fuel consumption.) By reducing the amount of batteries, hybrid vehicles let car designers exploit more fully the efficiencies permitted by lightweight materials and more aerodynamic designs. Like its battery-powered cousin, a hybrid-electric would recapture the energy of its own momentum through regenerative breaking.[21]

The Lovins "Supercar," a concept vehicle that has been designed and "tested" on a computer, would have a less-than-15 kilowatt (20 horsepower) engine (probably a small gas turbine or diesel) that powers highly efficient electric motors that are integrated into the wheel hubs. It would not require components that are routine in today's cars, including transmission, driveshaft, universal joints, and differential. The car would weigh less than 600 kilograms (1,300 pounds)—60 percent less than a conventional car—and have much better aerodynamics and more-efficient tires. The result, according to a computer model used by General Motors, is an automobile with a fuel economy of about 1.6 liters per 100 kilometers (150 miles per gallon).[22]

Much of the technology needed to build such a vehicle already exists—according to one analysis, the GM Impact could be converted to a hybrid that needs less than 2.3 liters per 100 kilometers, getting more than 100 miles per gallon—though the electronic integration of the technologies presents a challenge. Hybrids have yet to receive the kind of attention that pure electrics have, largely due to the ZEV requirement, but several companies, including Saab, Volkswagen, and Volvo, are developing hybrid vehicles. Nevcor, a California company, is designing a system for quickly converting existing car models to hybrid operation. Ford and GM are both working on advanced hybrid vehicles in programs co-funded by the U.S. Department of Energy. Although these efforts are not well publicized—indeed, many of the details are considered top secret—those close to the engineering teams involved report rapid progress and a strong likelihood that revolutionary new vehicles will be unveiled soon. Meanwhile, data available from several experimental and concept vehicles demonstrate the large envrionmental gains that can be achieved with the new cars. (See Table 10–1.)[23]

The Swiss have taken another approach to bringing

TABLE 10-1. *Characteristics of*

Model	Fuel	Weight	Fuel Efficiency
		(kilograms)	(kilometers per gigajoule)
1975 Ford Torino[2]	Gasoline	2200	66
1994 Ford Taurus[2]	Gasoline	1600	137
1994 Honda Civic VX[2]	Gasoline	950	240
1994 Dodge Caravan[3]	Natural Gas	2200	89
Volvo ECC (Hybrid)[4]	Diesel	1580	178
GM Impact[5]	Electricity	1318	262[6]

[1]Excludes methane except for the Torino. [2]Emissions based on a 50,000-mile urban/suburban testing cycle. [3]Emissions based on 100,000-mile testing cycle; emissions at 50,000 miles are probably lower. [4]Emissions based on an urban driving cycle; emissions

SOURCE: See endnote 23.

electric cars to the world's roads. The Tour de Sol spurred small-scale inventors, with modest government support, to pioneer a new generation of two-passenger, battery-powered city cars made of lightweight fiberglass. They typically have a range of 50–80 kilometers and a top speed of 50–100 kilometers (30–60 miles) per hour. While these cars run on conventional electric motors and lead-acid batteries, the Swiss engineers who developed them have avoided becoming obsessed with the batteries, and instead developed a car that meets everyday transportation needs by adapting available technology. Even in the United States, surveys show that 70

Selected Automobiles

		Emissions	
Hydro-carbons[1]	Carbon Monoxide	Nitrogen Oxides	Carbon Dioxide
	(grams per kilometer)		
0.62	3.11	1.68	365
0.05	0.36	0.11	176
0.05	0.17	0.10	100
0.01	0.22	0.01	195
0.01	0.08	0.11	140
0.00	0.03	0.02	51

based on an urban/suburban testing cycle would probably be lower. [5]Emissions based on current California electricity generation mix, which is taken to be representative of future marginal electricity supply in the United States and elsewhere. [6]Includes an estimated 67 percent power generation and distribution losses.

percent of drivers average less than 80 kilometers (50 miles) a day. A 2,000-kilogram car with a range of 600 kilometers is not necessary for such trips.[24]

From the outside, the small Swiss electric cars appear tiny, but inside they have ample room to seat two passengers and a full load of groceries. They weigh less than half as much as a normal car, accelerate rapidly, and are so short that the driver can skip parallel parking and just turn directly into the curb. Horlacher AG, a small, family-run Swiss composites company, has built prototype cars using honeycomb fiberglass bodies that emerge from a mold rather than from metal fabrication machines. The Horlacher cars are less than three meters (nine feet) long, weigh 400–600 kilograms, and have a range of 150–400 kilometers (90–250 miles), depending on the batteries used. Several other small companies have similar efforts under way.[25]

Urs Muntwyler, the Swiss engineer who cofounded the Tour de Sol, says he is determined not only to reinvent the automobile but to "transform modern transportation." The goal is to provide small, emission-free city cars, and at the same time to integrate light electric vehicles (LEVs) with more-efficient transport modes such as mass transit and intercity rail. The Swiss are strong supporters of public transportation, and their rail system is widely acknowledged to be among the world's best; Swiss planners are already adapting rail cars so that they can be used to carry LEVs from town to town.[26]

Throughout Europe, and to a lesser extent in Japan and the United States, the idea of using small, battery-powered electric cars in cities is gaining ground. The Italian company Fiat is already in the showroom with electric versions of its popular Panda and Cinquecento subcompacts, and the French automaker Renault has built an electric commuter car called the Zoom that is two meters long and has a range of 150 kilometers. In the United States, the Sacramento Municipal Utility District is experimenting with several small European electrics, and is considering a program to subsidize customers' purchases of such cars. Experts differ vigorously over how large this niche market might become, a question that hinges in part on changing peoples' habits and in part on how much batteries are improved and how much less expensive than conventional cars the small electrics eventually become.[27]

The environmental advantages of electric and hybrid

vehicles need not be limited to passenger cars. In the past, many city buses were powered by overhead electric lines, and several manufacturers are now working on battery-powered and hybrid buses and delivery vans. These larger vehicles are well suited to electric vehicle technology. They have plenty of space to hold batteries or other necessary equipment, and because they are usually driven along set urban routes, recharging presents fewer problems. Driving in another direction, Paul MacCready has proposed taking the Swiss LEV concept a step further, developing tiny electric "subcars" for short, low-speed urban trips and battery-assisted bicycles that would bring the advantages of convenient two-wheeled travel to more people. Several electric bikes have already been built.[28]

* * * *

One of the big advantages of a shift to efficient hybrid-electric vehicles is that they open up an array of new fuel possibilities. A main limitation of gaseous fuels such as natural gas and hydrogen is that they require relatively heavy and bulky storage tanks. But a hybrid might be able to go 700 kilometers (450 miles) on the equivalent of just 11 liters (three gallons) of gas. Assuming that such a vehicle were powered by natural gas rather than gasoline, a similar range could be achieved with a three-foot-long, 13-kilogram (29-pound) pressurized natural gas tank that was built and certified for automotive use in 1993. Such cars would emit 85 percent less carbon dioxide than today's automobiles, and 95 percent less carbon monoxide and nitrogen oxides. Indeed, a well-designed hybrid would have lower emissions than an equivalent battery-powered car that is recharged from the average mix of power plants in use today.[29]

Given the array of new technologies, it is hard to predict the mix of fuels and vehicle types that will emerge. In all probability, it will consist of battery-powered cars for local use and hybrid vehicles for longer trips. In either case, the way would be open for natural gas and electricity to begin displacing oil-based fuels. So long as the fuels are used efficiently, their use would increase only modestly. Indeed, if half the U.S. passenger car fleet were converted to efficient natural-gas-powered hybrids, less than a 7-percent increase in U.S. gas supplies would be needed to fuel them. If the other half were converted to battery-powered electrics, less than a 10-percent increase in the country's electric power supply would be required.[30]

Most cities already have electricity and natural gas distribution systems, making it relatively easy to provide refueling. Battery-powered cars, for example, may be recharged at home at night from the local power system, using "off-peak" electricity. Eventually, parking lots might also be equipped with rechargers, allowing cars to be topped off while their owners were at work or shopping. And quick-charging devices could be available at service stations for emergencies. Natural gas refueling can also occur at home, but service stations will probably continue to be the more common source. In Japan, the government is already planning a national network of Eco Stations that will offer compressed natural gas as well as electricity.[31]

In the longer run, hybrid technology could give an even greater boost to hydrogen fuel, which has the disadvantages of being bulky to store and expensive. As a result, hydrogen can only become an attractive automotive fuel if it is used very efficiently. One of the most promising long-term options is to use it in a fuel cell engine, similar to those in spacecraft, that produces electricity at an efficiency of 35–65 percent. Although fuel-cell-powered cars and buses are already being developed, the technologies will need to be improved in order to be economical. In any case, both hydrogen and electricity open up the possibility of one day powering cars, trucks, and buses without using any fossil fuels at all.[32]

Regardless of the mix of autos, buses, electrics, and hybrids that emerges, better storage devices will be needed. Most of today's electric cars rely on lead-acid batteries, chemical devices that carry an electric charge. Although such batteries have been used to start automobile engines since early this century, their high cost, short lifetime, and limited storage capacity are a drawback—even for hybrids. Since 1990, this has stimulated extensive efforts to develop a better battery, including the Advanced Battery Consortium (a four-year, $260-million joint development program of the U.S. government and the auto industry) and a similar effort in Japan.[33]

Researchers see a potential to increase the energy storage capacity of lead-acid batteries by 50 percent or more, while other research teams are seeking to adapt the more efficient but costlier nickel-cadmium and lithium batteries used in computers and other electronic devices. Finally, a number of researchers are working on a range of new battery materials, from sodium-sulfur to zinc-bromine, nickel-chloride, and nickel-metal-hydride. A 1994 study by the California Air Resources Board suggests that battery technology continues to advance rapidly, and that by the end of this decade new batteries may double the 100–150 kilometer range typical of today's electric cars. In the longer run, even lower costs and greater storage capacities are within reach.[34]

One of the most promising storage devices now on the drawing board is the flywheel, a mechanical battery that stores energy in the form of a spinning disk rather than in chemical form. Simple in concept, metal flywheels have found limited application for decades, but only became practical for motor vehicle use with the development of strong, lightweight composite materials that can be spun in a vacuum at up to 200,000 revolutions per minute. By the early nineties, several companies were working on advanced flywheels with electromagnetic bearings, an internal motor/generator, and the potential to store and release energy at an efficiency of more than 90 percent. The ability of flywheels to deliver large bursts of energy makes them particularly well suited to hybrid cars. And unlike most batteries, which would need to be replaced several times during the life of a vehicle, flywheels would probably outlast most cars.[35]

The father of the modern flywheel, Richard Post, an 80-year-old theoretical physicist at California's Lawrence Livermore Laboratory, believes that the relative simplicity of this technology will make it possible to build flywheel-powered cars by 2000 and to rapidly scale up production and lower costs soon thereafter. Flywheels may have even more appeal as a supplementary power source on hybrid buses, trucks, and locomotives.[36]

★   ★   ★   ★

The emergence of fundamentally new transportation technologies for the first time in several decades presents new challenges for the dozen or so automakers that dominate the business today. One of the big questions is how long the transition to lightweight electric vehicles might take. History provides one guide. In the early years of the twentieth century, it took gasoline-powered engines less than a decade to eliminate what had been vigorous competitors. Then between 1920 and 1926, automobiles went from being 85 percent wood to 70 percent steel. Although conventional wisdom says that the much larger scale of today's industry makes rapid change impossible, recent experience in the computer industry and other fields suggests otherwise. If the new technologies prove themselves economically and environmentally superior to today's vehicles, they may push the old technology aside earlier than most experts expect.[37]

The possibility of rapid change, combined with confusion about the precise mix of new technologies, has left many executives understandably anxious. In the United States, that anxiety is mixed with belligerence, bred of a concern that company statements about the potential to build more-efficient, less-polluting vehicles may lead to a government regulation requiring them to do so. As a result, U.S. auto companies let their lawyers and public relations firms speak for them, regularly attacking the ZEV standard in court and before the state legislature, while proclaiming boldly that it will be decades before they can build an affordable electric car.[38]

In Europe and Japan, by contrast, the major auto companies are more publicly enthusiastic about their new investments. European manufacturers that have built experimental electric cars include BMW, Mercedes, Opel, and Peugeot-Citroen, while Volkswagen and Volvo are among those working on hybrids. Japanese auto companies, which since the fifties have based their R&D programs on U.S. regulatory standards, have

taken the California zero-emission requirement particularly seriously. Nissan, for example, is working on a
four-door electric car that will have a top speed of 90
kilometers (55 miles) per hour and a range of 120 kilometers (75 miles). Like Toyota, Isuzu, and Mazda,
which are also developing electric vehicles, Nissan will
find a guaranteed market for its first sales: the fleets of
local governments.[39]

At the same time, the expanding market for cleaner
vehicles has lured a growing number of small entrepreneurs, some of whom see themselves as the next Henry
Ford. In Massachusetts, for instance, the Solectria
Company—headed by James Worden, a 26-year-old
self-described electric car nut—is converting conventional cars into electrics and marketing them to eager
buyers. A similar venture in California, called U.S. Electricar, has converted 200 Geo Prizms into electric cars
with a range of 130 kilometers. Both companies have
plans to build such vehicles from the ground up in the
near future. Swatch, the Swiss watch manufacturer, has
developed a stylish electric car that it plans to manufacture in cooperation with Mercedes Benz.[40]

The new market is also proving irresistible to aerospace and electronics firms looking for new business in
the aftermath of the cold war. In the United States,
companies such as Allied Signal, Lockheed, and
Motorola have begun to invest in components for electric and hybrid cars. Motorola estimates that each such
vehicle may include $800 worth of electronics, with an
offsetting reduction in such devices as transmissions,
carburetors, and mechanical drive trains. In California,
a consortium of high-tech companies has banded together with various government agencies to form an organization called Calstart in order to facilitate entry into

the automotive industry. Relying on the products of its members, Calstart built a state-of-the-art experimental car in 1993.[41]

Since much of the expertise in new technologies lies outside the traditional car industry in a mixture of large and small companies in the electronics, aerospace, chemical, and other industries, the large automotive companies will need to form partnerships in order to move forward, something that many are already doing. Still, auto company executives are wary of the possibility that their existing facilities—used to produce piston engines, stamp steel, and assemble the final product—may shortly be transformed into multibillion-dollar tax writeoffs. Like AT&T before the breakup of the U.S. telephone monopoly or IBM before the advent of the personal computer, the large automakers are reluctant to give up the comforts of a relatively staid technological marketplace.[42]

Still, behind the scenes, many seem to understand the inevitability of the coming revolution. Like the little Dutch boy with his finger in the dike, the auto industry is unlikely to hold back the tide of technological change for long—even if it wants to. Indeed, the possibility of a shakeout looms large. New market opportunities may bring new players into the auto business while some of today's dominant companies go under. Moreover, if cars become more like computers, evolving rapidly and unpredictably from year-to-year, the business could present real difficulties for the corporate cultures of General Motors, Toyota, or Volkswagen.

Today's auto executives would do well to remember that most buggymakers never made the shift to automobiles, and that Western Union did not make the transition from the once-lucrative telegraph business to the far

larger telephone industry. In the current globally competitive auto business, the market is unlikely to be kind
to those who fall behind the technological curve.

★   ★   ★   ★

No matter how much less polluting automobiles
become in the future, one thing is clear: they will not be
a panacea for the world's transportation problems. Although the new technologies could greatly reduce many
of the energy-related problems created by cars, they
could exacerbate others, including the suburban sprawl,
congestion, and destruction of neighborhoods that is
rampant in so many parts of the world. This suggests
that the redesign of the automobile must be accompanied by efforts to spur an array of new transportation
options and to change regional development patterns so
as to reduce the need for travel and create more livable
communities.

The tremendous growth in the number of automobiles in recent decades has largely obviated the freedom
and convenience that the car was originally supposed to
provide. Today, millions of people reluctantly spend
several hours in a car each day just getting to and from
work. Moreover, in many countries the rapid spread of
roads and dispersed housing units is eating up cropland
that is needed to grow food.

The ultimate answer to the problem of endless sprawl
and unlivable cities lies in more environmentally sustainable patterns of development. (See Chapter 11). At
the same time, a critical short-term priority is to step up
investment in public transportation and other transport
options. These, too, are becoming more diverse and
more attractive as a result of recent developments. And
unlike the automobile, which is only within the economic reach of an estimated 10 percent of the world's

population, public transportation can benefit the majority.[43]

Efforts to forge a more balanced transportation system will be impeded by the fact that the playing field has been severely tilted at least since the late forties, when the U.S. auto industry purchased and then destroyed many of southern California's rail lines and bus systems in order to spur automobile purchases. In recent decades, U.S. subsidies to the automobile and truck—principally for road-building—have greatly outstripped those for rail and bus systems. Yet transportation planners acknowledge that public transportation is far more efficient than driving a car, with commensurate reductions in air pollution. Moreover, by adapting the new technologies just discussed to trains, subways, and buses, the environmental benefits of public transport can be multiplied.[44]

High-speed intercity rail links already cover most of the heavily populated areas of Japan, and similar systems are being built in several parts of the United States, including California and Texas. In Europe, 12 countries are working on a $76-billion plan to link major cities with 30,000 kilometers of high-speed rail lines. Already, it is quicker to use rail than a jet to travel between some European cities that are up to 700 kilometers apart, if the time it takes to get to the airport is included. Within cities, many governments have recently stepped up their commitment to light rail commuter lines and to urban bus systems. In California and other parts of the United States, many of the air pollution reduction plans being developed in response to the Clean Air Act amendments of 1990 include extensive efforts to promote reliance on public transportation and on ride-sharing and vanpools.[45]

Another promising and pollution-free alternative to

the automobile that is generally overlooked by transportation planners is the bicycle. Close to 1 billion bicycles are thought to be in use worldwide, far surpassing the number of cars. Although in industrial countries many are used primarily for recreation, they are a vital transportation technology in many developing nations, where most people cannot afford a car.[46]

Today, bicycles are used extensively for commuting in China and India, for hauling produce in Kenya and Indonesia, and for delivering mail in Tanzania. For such countries, adding large numbers of cars to the roads (as most are planning) threatens to impede transportation for most people while aggravating already dangerous levels of air pollution. In countries such as China, population density alone makes it inconceivable that pursuing a U.S.-style, auto-centered transportation system (even with ZEVs) would bring anything but grief to most people—worsening congestion and paving over land needed to grow food. If governments are to avoid becoming trapped with unsustainable transportation systems, they will need to encourage more rather than less reliance on the bicycle.[47]

★   ★   ★   ★

With so many opportunities to develop new transportation options as well as new kinds of motor vehicles, one of the most pressing issues is how to spur the transition. Perhaps the most efficient way to do this is to incorporate the environmental costs of various transportation options directly in the prices people pay for them—in gasoline prices, licensing fees, road tolls, and so forth. Reducing the multibillion-dollar subsidies to road travel is one obvious step. Another is to increase fuel taxes, particularly given the fact that wholesale gasoline prices were at near record low levels in the mid-nineties.[48]

The high gasoline and automobile taxes already in place in Europe are one approach to remedying this problem, as are proposals to incorporate such costs in the insurance or annual licensing fees that automobile owners must pay in most countries. Some analysts suggest taking this a step further, taxing cars differentially, depending on the amount of pollution they produce. If such taxes were set high enough, they would discourage heavy reliance on the automobile and at the same time encourage manufacturers to produce cleaner cars.[49]

In the absence of full price reform, financial incentives can also give electric and hybrid cars a boost. By eliminating vehicle taxes for electrics (as Switzerland has done), or adopting a range of other pricing schemes, governments can encourage drivers to buy the new vehicles. Austria, Germany, and the Netherlands have adopted modest financial incentives for low-emission vehicles, and California is considering a gas guzzler tax for polluting cars while giving a rebate for zero-emission vehicles. In Sweden, the government is planning a bulk purchase of electric cars, encouraging manufacturers to scale up production.[50]

Although market-based incentives are more efficient in theory, the public (through its elected officials) often prefers a regulatory approach to cleaning up the world's transportation mess—ranging from automotive emission limits to fuel economy standards. California's ZEV standard is a classic example. It is narrow and inflexible, but it is popular with California voters. And according to a review by state government officials in 1994, it is accomplishing the intended goal of accelerating the development of electric vehicles. Switzerland took a similar approach to promoting rail transportation when in

1994 voters decided to ban all trans-Alpine truck traffic as of 2004, a step that may help shift the entire European freight system toward rail.[51]

Mandates clearly have a role to play, though they would be more effective if they were accompanied by financial inducements to purchase the new vehicles. In California, several electric utilities have applied to the state government to subsidize their customers' purchases of electric cars.[52]

In the end, the only real solution to the world's transportation energy problems is to attack them with multiple policies on many fronts. Efforts to produce cleaner vehicles can go only so far, while efforts to accelerate reliance on other transportation modes will reduce, not eliminate, reliance on the car. A more sustainable transportation system ultimately will be a more diverse one, relying in part on cleaner cars but also moving to much heavier use of public transportation and bicycles. If the goal is a higher quality of life, the challenge can only be met if we create a more livable urban environment—which should not be defined by air quality alone.

11

Building for the Future

In 1978, the board of what is now the Internationale Nederlanden (ING) Bank decided it needed a distinctive new image to revive its flagging business, and what better way to start than to build a new headquarters. With environmental issues already near the top of the Dutch agenda, the directors decided not to imitate the stodgy steel-and-glass office buildings of its competitors, but rather to build an "organic" building that integrated natural shapes, low noise levels, green plants, and art in something that celebrated the human spirit. Conceived of in the midst of the Iranian oil price shock, the ING Bank building was also designed to be highly efficient in its use of energy and other natural resources.[1]

The results were more dramatic than most architects and developers thought possible. When completed in

1987, the new headquarters used 80 percent less energy than a contemporary neighboring office building, and more than 90 percent less than the bank's previous headquarters. Among the measures that permitted such reductions in energy bills were insulation materials in the walls and optimal use of natural light, a practice known as daylighting. (No desk is more than six meters from a window, for example.) The windows also allow solar energy to provide much of the building's heat in winter, with a natural-gas-burning generator providing the remainder plus most of the building's electricity. Instead of relying on conventional air conditioning, the building uses natural and mechanical ventilation, backed up by an absorption cooling system run by waste heat from the cogeneration plant.[2]

The additional costs incurred during construction of the ING Bank were covered by lower energy expenditures in just its first four months of operation. What surprised skeptics, however, was not the reduction in energy costs—those had been expected—but the 15-percent drop in employee absenteeism and the gains in worker productivity. Just as important to the bank's managers was the boost in public image and the resulting business the company experienced when it moved into its new home. By the early nineties, the ING Bank headquarters was the best known building in the Netherlands.[3]

Though somewhat more innovative than most, the energy-conserving ING Bank represents a generation of more efficient and less environmentally damaging buildings that have begun to appear in the world's cities and countryside over the past two decades. From single-family homes in Canada to towering office buildings in Singapore, energy use has been slashed to 30–90 percent below earlier levels. And some innovators want to go even further.

New York City architect William McDonough, a pioneer of "environmentally friendly" architecture, believes that the new wave of buildings represents a reversal of trends under way since mid-century. "With the advent of cheap energy and the large sheet of glass," McDonough observes, "architects and designers more often used physical, mechanical, and chemical technologies as brute forces to subvert nature rather than to harness its readily available power without depleting its non-renewable resources." If McDonough and other innovators have their way, mother nature may once again become the leading energy supplier for the world's buildings.[4]

* * * *

Buildings, along with transportation and industry, are one of the three largest claimants on the world energy system. Heating, cooling, and lighting the world's built structures suck up roughly one third of the massive flows of energy used by modern societies. Yet these claims can be greatly reduced through relatively modest changes in designs and materials. From a technical standpoint, this is the part of the world energy dilemma that may be the easiest to solve; from the institutional point of view, it may the hardest.[5]

Some headway has already been made. The amount of external heating required in an average home in the United States was cut by more than 40 percent between 1973 and 1990, while in Denmark, the reduction was more than 46 percent. These gains stem in part from more-efficient technologies—double-pane windows, thicker insulation, and better furnaces, for example—as well as tighter construction techniques and more-efficient ventilation. The most dramatic gains have been made in new buildings, but extensive retrofits have also played a role. During the past two decades, millions of old buildings have had weather-stripping added around their windows and extra insulation put in their attics. Nonetheless, large additional gains are still possible.[6]

Small homes and large office or apartment buildings have a basic difference that affects the way they use energy—as well as strategies for reducing that use. In temperate climates, small homes typically use most of their energy for space and water heating; conservation strategies that focus on tight construction, insulation, and improved windows tend to be most effective. Larger buildings, on the other hand, have more artificial lighting and a large internal heat load, and typically use as much as two thirds of their energy for lighting, cooling, and ventilation. Indeed, many such buildings are designed so poorly that they must be cooled even when outdoor temperatures are comfortable. Nor is it uncommon for the cooling system to be on in one sector of such a building while the heat is on in another.

One of the priorities in reducing energy use in large buildings is improved lighting. The newer lights discussed in Chapter 4 can cut the energy used for illumination by 50 percent or more and can lower the amount of unwanted heat produced by those lights. A second priority, generally limited to new buildings, is to use advanced double-pane windows. Many have a special low-emissivity coating (a heat mirror) that uses an atoms-thin layer of silver to filter out infrared rays but allows visible light to enter, further cutting the amount of heat that accumulates in the building. Along with improved lights, these new glazings, which insulate four times as effectively as a single pane of uncoated glass, have eased the work of energy-conscious building designers. In fact, once the external and internal heat load is reduced, engineers often find they can make the building's air-conditioning systems smaller, reducing the initial cost of construction.[7]

The Pacific Gas and Electric Company (PG&E), a large California utility, demonstrated the potential for retrofitting commercial buildings in 1993. Responding to a challenge by energy efficiency experts working for environmental groups—who claimed that a 75-percent cut in energy use was economically justified—PG&E sought bids from several engineering firms for a retrofit of its research headquarters in San Ramon. When completed, the retrofit achieved an immediate 55-percent reduction in electricity use through more-efficient lighting, better windows, and a smaller cooling system. The total reduction is expected to reach 72 percent—just short of the original goal—when the building's energy-guzzling office equipment is replaced.[8]

The potential gains are much greater in central and eastern Europe, which are renowned for the shoddiness of construction and maintenance practices during the communist era. In Russia, for example, where most buildings do not have thermostats, windows are designed for easy opening so that people can cool overheated apartments in mid-winter. Installing simple boiler controls, thermostats, and meters could lower the amount of energy used for indoor heating by up to 45 percent, with the cost of the measures being paid back in saved natural gas in just a year and a half, according to one study. These energy levels could be halved again by adopting some of the lighting and heating measures now standard in western buildings.[9]

As in many areas of energy development, poor countries have trailed industrial ones in improving building design and construction. But with high-rise office and apartment buildings now crowding their skylines, Third World cities no longer can afford this waste. Like large buildings everywhere, these structures require far more energy for lighting and cooling than for heating, a problem exacerbated by the hot, humid climates common in much of the developing world. Yet most are still built with inefficient lighting systems and windows made of a single sheet of plain glass. In Thailand, a remarkable 40 percent of the projected growth in electricity demand will come from new commercial buildings. This trend can only be turned around if the new building innovations being pioneered in industrial nations are also adapted by—and adopted in—developing countries.[10]

For more-efficient detached homes, one of the most exciting innovations since the seventies is the superinsulated house, first developed in Sweden. During the past decade, roughly 100,000 such homes have been built in Scandinavia, Canada, and the United States. They typically combine high-quality, airtight construction with abundant insulation and double- or triple-glazed windows, sometimes filled with inert, highly insulating gases. Such homes virtually warm themselves through the waste heat from the occupants, lights, and appliances. Some new Canadian models (known as R-2000 homes) are so airtight that they require special air-to-air heat exchangers to provide ventilation without wasting heat. Engineers point out that with such systems installed, R-2000 homes often have cleaner air than conventionally built ones.[11]

Similar designs also work in hot climates. In California's blistering Central Valley, where summer temperatures can reach 40 degrees Celsius (105 degrees Fahrenheit), one custom-built home has eliminated both the furnace and the air conditioner by relying on extra insulation in walls and doors, a ceramic-tile floor to increase thermal mass, a whole-house fan for ventilation, and other measures. The Davis house, which is part of PG&E's effort to push the limits of efficiency, is expected to achieve a level of energy use that is 65 percent below California's already tough 1993 building code—at a projected commercial cost that is actually less than that of a conventional home.[12]

★　★　★　★

Beyond the fundamentals of more-efficient design and construction, pioneering architects and urban planners are also considering the overall place of buildings in the natural environment—from the materials they are built with to the way they use the sunlight that strikes them. In fact, studies show that it takes large amounts of energy to extract, refine, and fabricate the materials used to construct buildings. This so-called embodied energy, which typically is the equivalent of 5–10 years' worth of energy used to operate the building, can be reduced by minimizing the amount of materials needed or by switching to less energy-intensive ones. The Davis home, for example, required about half the normal amount of lumber.[13]

Choice of building materials can also affect energy use in other ways. Lighter-colored materials, especially for the roof—which receives much more sun in summer when it is not wanted than in winter when it is—can cut peak cooling needs by as much as 40 percent. Similar improvements can be achieved by placing vegetation, including large shade trees, around buildings. Deciduous trees block unwelcome summer sunlight without impeding solar heating in winter. Both methods also contribute to reducing energy demand by minimizing the "urban heat island," the boosting of urban temperatures when pavement and building surfaces absorb the sun's rays instead of reflecting them. Researchers from the Lawrence Berkeley Laboratory have found that cooling needs can be cut 30 percent by planting enough trees.[14]

More effective use of sunlight—both for heating and lighting—has also attracted the attention of architects. Selectively glazed windows, skylights, atria, reflectors, and even high-tech systems that "pipe" sunlight deep inside are cutting the need for artificial lights in many new buildings. Steven Ternoey, a Colorado-based daylighting pioneer, notes that "even with advanced lighting technologies, nothing is more efficient or more pleasing to the eye than using natural light to illuminate a building space." For schools and offices that are occupied primarily during the daytime, daylighting is particularly cost-effective in cutting peak electricity demand. At the same time, it improves lighting quality, aesthetics, and worker productivity—benefits that Ternoey and others believe outrank the energy gains.[15]

Passive solar heating is another important way for architects to use sunlight in their designs. Although it has been practiced in traditional architecture for thousands of years, passive solar building design experienced a renaissance in the mid-seventies. Designers discovered that these buildings need not be unsightly mazes of glass and stone. Simple orientation of the home can boost the solar gain by locating most windows toward the equator, while cooling needs are lowered by reducing the amount of glass on the west to avoid afternoon sun. Though most passive solar homes are well insulated and have "thermal mass"—usually masonry or rocks built into them to retain heat at night—they blend solar principles into a range of traditional and modern architectural styles. No firm numbers are available, but at least a million passive solar homes probably exist today, even though they are currently less popular than in the seventies.

Passive solar homes have found their main niche so far among architect-designed, custom-built homes. One builder attempting to take them into the mainstream is Paul Neuffer of Reno, Nevada. He is building passive solar homes that use half the natural gas of a typical home in the area. The extra costs add roughly 1 percent to the price of the home—less than the local utility's cost of providing the amount of gas that is saved. Under a new conservation program, the utility is proposing to give rebates for the construction of such homes—saving money for ratepayers as well as the new home buyers. According to scientists at the National Renewable Energy Laboratory in Boulder, Colorado, passive solar design can reduce the amount of heating energy used by residential buildings by as much as 70 percent, while commercial building energy needs can be lowered by 60 percent.[16]

In 1992, the Canadian Energy Ministry launched a design competition to double the efficiency of the already efficient R-2000 homes. Ten designs were selected and built, with expected energy needs cut by as much as 75 percent. In addition to the insulation, efficient appliances, and triple-glazed windows generally found in R-2000 homes, solar orientation met half the remaining heating needs in half the superefficient homes examined, demonstrating the extensive synergies between conservation and solar heating strategies. The Canadian homes also disprove a common misconception: that solar buildings are only for sunny regions such as Arizona or Tunisia. Even in the overcast climates of Denmark and New England, passive solar design can meet a significant share of a building's heating needs.[17]

One of the most clever recent developments is using a building's structure not just for collecting solar heat but for generating electricity as well. Integrating solar photovoltaic (PV) cells (see Chapter 8) into a building's roof or facade can make structures serve dual purposes. At the household level, most residential homes equipped with solar cells use reversible meters instead of batteries, meaning that excess power is transmitted to the power company's grid at the same price that the resident purchases power. When the sun is not shining, the building relies on electricity provided by the local utility.

Steven Strong, an architect who once worked on the Alaska oil pipeline, is probably the world's leading pioneer in solar-powered buildings. In 1979, Strong, who heads a Massachusetts company called Solar Design Associates, designed the world's first grid-connected PV home, in the Boston suburbs. This large, upscale home includes 7.5 kilowatts of polycrystalline silicon solar cells that provide about as much electricity each year as the house uses—though not always at the precise times it is needed. Strong has since worked with the New England Electric System to install 38 rooftop systems in Gardner, Massachusetts, and with the Sacramento Municipal Utility District on an even larger effort in California.[18]

By the mid-nineties, several companies were at work on techniques for incorporating solar cells right into the facade glass normally used in commercial and institutional buildings. The large German glass manufacturer Flachglas (which also makes the reflective glazing for the Luz solar thermal plants described in Chapter 7) is integrating polycrystalline solar cells into a semitransparent glazing it is now testing. The result is a solar-electric window that provides filtered light for occupants while at the same time generating electricity. By the end of 1993, Flachglas had installed several prototypes of its system in commercial buildings in Aachen, Hannover, Berlin, and four other cities. Across the Atlantic, Siemens Solar and a U.S. firm, Corning, are working to incorporate thin-film solar cells into manufactured glazing. Meanwhile, PV companies in Japan, Switzerland, and the United States are testing new types of solar cells that also function as roofing shingles.[19]

Experts believe that as solar cell costs drop, solar-electric glazing of office buildings' facades may become commonplace—particularly in high-latitude countries where the sun is usually low in the sky. Since heavy energy use in commercial buildings often coincides with periods of strong sunshine, the match is an especially good one. Although the most popular initial market will likely be new buildings, these systems can also be economically retrofitted onto many older buildings. In the United Kingdom, the potential resource is estimated at 68 gigawatts—large enough to generate nearly half the country's 1993 electricity supply. A study of western Germany found a rooftop PV potential of some 50 gigawatts, a capacity that would account for one quarter of the region's current power output.[20]

As buildings become more efficient, and as they rely on external sunlight for many of their internal energy needs, it will also become more practical and economi-

cal for them to produce much of their own electricity—
by relying on small natural-gas-burning engines and fuel
cells as well as photovoltaics. (See Chapter 5.) Although
these technologies could in theory lay the foundation for
buildings that are entirely independent of the power
grid, this is unlikely in areas where a grid is already in
place. It is more likely that future owners of buildings
will both buy power from the local utility and sell power
back to it, depending on the time of day and the utility's
need for power. (See Chapter 12.)

★   ★   ★   ★

Changes like these will not come easily, however. Unlike
the automotive industry, the buildings sector in most
countries is highly decentralized not only horizontally,
but vertically as well. One study found as many as 25
diverse actors—from developers and mechanical engineers to bankers and maintenance staff—can be involved in the design, construction, and operation of a
single office building. Each makes decisions based on
personal economic motivations, often ignoring the
broader goal of keeping the building owner or occu-
pant's "life-cycle" costs to a minimum.[21]

The challenge is to make the industry as a whole more
like the pioneers who already sell buildings based in part
on low energy bills. Bigelow Homes in Chicago, for example, guarantees homeowners a rebate if their winter
heating bills are above $200 during the first three years.
It has only had to make good on its promise four times
since 1985. Likewise, whole developments in communities such as Santa Fe, New Mexico, and Davis, California, tout their use of sunlight to provide heating and
cooling.[22]

Nonprofit building associations, which control large

blocks of relatively inexpensive apartments in Europe and Latin America, have played a role in encouraging more-efficient buildings. KBI, a large Danish association, has begun retrofitting old apartments with passive and active solar features as well as energy and water-saving measures. KBI adds another floor to the three-floor apartments, as well as ample south-facing windows (converting the porch to a sun space in the process). The effort yields a 25-percent gain in floor space (at two thirds the cost of what constructing new apartments would be), while maintaining previous levels of energy and water use.[23]

Unfortunately, many deeply ingrained obstacles stand in the way of similar projects. These include a myopic focus on initial costs, obsolete rules-of-thumb for sizing equipment, outdated building codes, and taxes that target energy-saving equipment but not energy. In the past, most efforts to encourage more-efficient buildings have focused on tougher building standards. Although these have had positive effects in many countries, real progress will require concerted efforts to overcome myriad obstacles. The key actors are financial institutions, government agencies, builders, occupants, and owners.[24]

Several countries are now experimenting with environmental rating systems intended to ease buyers' choices and to spur the industry's adoption of new designs. The U.K. government started grading buildings on their environmental impact in 1991 using a system that penalizes the use of toxic materials in buildings and that allocates about two thirds of the rating points based on energy-related criteria. The program is so successful that British real estate agents include claims about high environmental ratings in their promotional materials.[25]

A similar effort has been launched in British Columbia, and in the United States, the nonprofit Green Building Council is exploring a related system. Austin, Texas, already has a model citywide program to promote green building construction—partly through a simplified rating system. In addition, the U.S. Environmental Protection Agency (EPA) has begun an Energy Star Building program that puts a green seal of approval on buildings that meet a range of energy and other goals. EPA's aim is to reduce the energy consumption of participating buildings by at least half.[26]

Government agencies and lending institutions can also encourage change by publicizing the fact that more-efficient buildings often have more usable space. By lowering the cooling needs of a building, air-handling ducts and blowers can be reduced, allowing designers to downsize the hidden, wasted space between floors. So much space can be saved that for each four stories, a fifth one can be added in a highly efficient structure, giving the developer more space to rent or sell, and making the project more profitable.[27]

Such arguments are likely to be most convincing to major real estate investors, including insurance companies and pension funds. Building tenants are generally more focused on the quality of their workspace and on what it might do to the productivity of employees. This is hardly surprising. For a typical business, employees' salaries cost roughly 160 times as much as the operation of the space conditioning systems. In other words, a boost of even 1 percent in worker productivity is worth more to a company than eliminating its energy bills entirely.[28]

Fortuitously, energy-smart buildings produce more productive employees who are more likely to show up for work. A recent review of 15 buildings (including the ING Bank) by the Rocky Mountain Institute identified a 5–15 percent gain in the productivity of people who work in "green" buildings. If employers begin to understand this equation, the dawdling pace of reform in the building industry may turn into a stampede.[29]

Gaining the support of financial institutions is particularly crucial. Since most buildings are constructed or purchased with borrowed money, banks, insurance companies, and other lenders have enormous leverage. Traditionally, lenders have treated innovation, whether for energy or other purposes, with skepticism—and higher interest rates—since it increases the perceived risk of a project. But recently a number of banks, often prompted by governments, have realized that by lowering utility bills, energy-conscious building designs leave owners more money to repay loans, reducing the risk of default. A few have even started offering lower interest rates for energy-efficient homes. The Bank of Montreal in Canada, for example, cuts the interest rate on loans for R-2000 homes by a quarter point.[30]

A variety of financing schemes have been put together in other countries. The Swedish National Housing Board offers subsidized loans for energy-efficient homes, a program that is at least partially responsible for the remarkably high standards of the Swedish building industry. In the United States, energy-efficient mortgages have been available through federal and state lending agencies, as well as private banks, for more than a decade primarily as a means of raising the qualifying ratios used by lenders, a practice that can also make homeownership affordable for more families. Still, most U.S. financial institutions resist making more significant changes. An attempt by the Vermont Housing Finance

Agency, for instance, to encourage lenders to lower interest rates on efficient homes failed, largely due to institutional resistance.[31]

Electric utilities in North America are making more rapid progress. Some have started helping home buyers with the downpayments on new houses that are energy-efficient. Such projects are a new and innovative element of the demand-side management programs that many electric and gas utilities are using to reduce the number of new power plants they must build. (See Chapter 12.) Ontario Hydro, for example, is offering rebates to design teams to cut the energy use of new buildings, and provides additional money once savings are measured. New Brunswick Power also offers rebates for R-2000 homes, which is one reason they now represent one out of five of the new homes built in the province—the highest rate in Canada.[32]

For commercial buildings, which often are not owned by the occupants, other mechanisms are needed to promote efficiency. Among the provisions that might be included in standard leasing agreements are requirements that an energy audit be undertaken, that specific efficiency investments be made, that allocation of costs and benefits from such investments be fairly distributed, or that building systems be serviced and upgraded on a regular basis.[33]

★   ★   ★   ★

Buildings and transportation together account for more than half the world's energy use. Although opportunities for improving the efficiency of each of these sectors are enormous, realizing the full potential—and dealing with other issues that affect the quality of modern life, such as congestion and disappearing neighborhoods—can only be achieved by addressing the structure of cities themselves. The era of cheap energy in the middle of the twentieth century encouraged urban planners to become preoccupied with skyscrapers and motor vehicles. The cities they built separated people from the natural world and from each other. Righting this imbalance is a prerequisite to a sustainable global energy system.[34]

Sprawled, low-density developments were made possible by automobiles, and are now almost totally dependent on them. As a consequence, low-density U.S. and Australian cities use several times as much gasoline per person as the higher-density cities of Europe and Japan. If buildings continue being spread across the countryside, it may take as much energy transporting people and goods to them as is used in the buildings themselves. At the same time, such development patterns have lowered the quality of life for millions, forcing them to spend unwelcome hours in their automobiles each day.[35]

"High density" makes many people think of towering apartment buildings and limited open space, yet dense developments can be pleasant and lively if they are well planned and have plenty of green space. According to one study, a compact development can mix small apartment buildings and town houses with clustered single-family homes, and still leave 30 percent of the area for open space and parks. (In a typical low-density suburb today, only 9 percent of the land is devoted to open space.) Even passive solar buildings can be built as densely as 35–50 dwellings per hectare. A normal U.S. residential suburb, by comparison, is zoned for no more than 10 homes per hectare.[36]

The density of human settlements is shaped in large part by the decisions of local, regional, and national officials. In Europe, governments have a long history of using zoning, tax incentives, bans on low-density projects, and other measures to ensure that cities remain compact. Swedish cities—compact centers surrounded by vast stretches of rural, largely forested land—demonstrate the success of that nation's urban land use policies.[37]

Planners in large European cities such as Paris and Stockholm have encouraged high-density development patterns so as to maintain traditional charms as well as to save energy. They often do so by advocating a mix of land uses—so that jobs, homes, and services are not scattered far and wide. They strive to focus new developments within cycling or walking distance of public transit stops. Stockholm, for instance, is ringed with satellite communities of 25,000–50,000 people each that are linked by rail and highways; shops, apartments, and offices are clustered around train stations.[38]

Without changes in development patterns, efforts to reduce the energy wasted in buildings and transportation will continue to fight an uphill battle. Promising measures include encouraging residential development in city centers by, for example, tieing office development to a required minimum amount of space for homes. An alternative is to levy fees on developers whose projects will aggravate the imbalance, and then use the revenues to create jobs near residential areas and housing near business districts. The Southern California Association of Governments' regional transportation plan, a 20-year effort to address the region's transport problems, has adopted just such an approach.[39]

In the long run, changes in human behavior will be needed if we are to achieve truly sustainable cities. While such notions often sound utopian, growing dissatisfaction with current patterns of living and working may well lead to profound behavioral changes. Among the more revolutionary recent developments is the mushrooming of communications technologies, which are already freeing many workers from the need to visit a particular work site each day. Telecommuting may one day reduce the need for travel greatly—and perhaps encourage people to devote more attention to the broader sustainability of their local communities.

12

Reshaping the Power Industry

In the late eighties, the Sacramento Municipal Utility District (SMUD), a publicly owned electric utility that serves a mainly suburban population of less than a million, stood at a difficult crossroads. Two decades earlier, it had bet its future on a 900-megawatt nuclear reactor called Rancho Seco. On paper, this was the company's largest asset. But it turned out to be a lemon, suffering frequent and prolonged shutdowns—pushing the local price of electricity up and the utility to the edge of insolvency. For nearly a decade, SMUD's managers fought bitterly and inconclusively with local citizens over the plant's future, trading claims about its economics and safety. Finally, in 1989 SMUD's customers voted to shut Rancho Seco down and begin the search for alternatives.[1]

To steer the process of change, the utility's board brought in as general manager one of the country's true energy revolutionaries: S. David Freeman, a tough-talking Tennessean who had shaken up several energy bureaucracies in his long career. At the Tennessee Valley Authority in the seventies, Freeman cancelled 8 of 12 nuclear reactors under construction. Earlier, he had helped shape the energy agenda of the Carter administration by heading a Ford Foundation study that identified an energy-saving potential almost as large as the one found by Amory Lovins.[2]

Freeman wasted little time in turning his newest charge around. While temporarily relying on power purchased from other utilities in place of the closed nuclear plant, SMUD immediately began investing in a range of alternatives, none of which included the large thermal power plants that for several decades have been the industry's mainstay. Instead, SMUD is now investing a remarkable 8 percent of its gross revenues on improving the efficiency of its customers' lights, windows, and other equipment and on planting 500,000 shade trees. Later in the nineties, SMUD plans to buy power from four industrial cogeneration plants, invest in a 50-megawatt wind farm, continue to install solar-electric and solar hot water systems on many of its customers' rooftops, and help in the purchase of electric cars and recharging stations in order to provide customers with less-polluting means of transportation. To accomplish these goals, SMUD is also pursuing R&D projects designed to accelerate the commercialization of many technologies. "What we're doing here ought to give people the courage of their convictions," Freeman said in a 1993 interview. "There *is* life after nuclear power."[3]

Although the SMUD experiments are at an early

stage, already they have allowed the utility to bring power prices well below the level of neighboring utilities with nuclear plants, and have inspired the local community with a can-do spirit that is a far cry from the adversarial legal battles between many U.S. power companies and their customers. The significance of SMUD's innovations can be seen by quickly thumbing through the utility's visitor book. This moderately sized power company—the fifth largest in its own state—now entertains guests from every corner of the globe, and many of them leave inspired by the notion of an electric utility that is decentralized, high-tech, participatory, and competitive.

Changes along the lines SMUD is pursuing are crucial if the world is to make the shifts to more sustainable patterns of energy supply and use discussed in earlier chapters. Electric power may only supply about 13 percent of the world's end-use energy, but in terms of fuel consumed or pollution produced, it constitutes about a third of the global energy economy. The electric industry is also one of the world's largest, with annual revenues estimated at more than $800 billion—roughly twice the size of the auto industry. Although many experts assume that it will be impossible to turn such a behemoth around, the current period of ferment and confrontation may be laying the seeds for a very different kind of power industry in the decades ahead.[4]

*   *   *   *

Creating a more sustainable power industry will inevitably require grappling with its core structure: the vertically integrated monopolies that are dominant in most countries. Although this structure provided reliable power and declining costs for more than six decades, most utilities have become lethargic and technologically backward, while systems of public regulation and control have become inefficient and adversarial. It may be tempting to tinker around the edges of this system, but it is now clear that the system itself must be changed.

The current structures took time to evolve. Chicago alone was served by four dozen power companies in 1900; a few decades later, single utilities were serving whole cities, then regions, and finally nations. Often, the same public or private enterprise now owns the power plants, transmission lines, and local distribution wires. With monopoly control came public responsibility, and in the United States, state commissions were formed to regulate the industry and to determine—in the absence of an open market—a fair price for power. In other parts of the country and around the world, utilities were taken over by city or provincial governments, and eventually were nationalized in the United Kingdom, France, and elsewhere.[5]

As the industry's structure changed, the price of electricity plummeted—from $4 per kilowatt-hour for U.S. consumers in 1892 to 60¢ in 1930 and just 7¢ in 1970 (in 1993 dollars). Demand soared, in some countries doubling every decade, and the industry adopted a "grow-and-build" model in which ever-higher consumer demand was used to justify scaling up the technology, which in turn would bring down the price and attract more customers. The impressive improvements in power technology stemmed, however, from a narrow frontier of advances in turbine materials and design. By the sixties, this bag of tricks was nearly empty. Average plant efficiencies were levelling off at between 30 and 35 percent (meaning that two thirds of the energy in the fuel was still dissipated as waste heat rather than being converted to electricity). (See Figure 12–1.) Since then, steam-cycle plant efficiency has stayed roughly even, and in some countries new plants are now less reliable than their predecessors.[6]

These relatively minor technical problems were joined by more serious ones when governments began to push the utility industry into nuclear power in the sixties. The plants appeared similar enough to a coal burner (the heat generated by splitting atoms was substituted for the fossil fuels used to boil water), encouraging utilities to order scores of nuclear plants far larger than any that had been tested. By the seventies, the unique hazards of atomic energy, and the consequent efforts to regulate plants, were adding dramatically to the complexity and costs of the new facilities—a problem vastly complicated by events at Three Mile Island

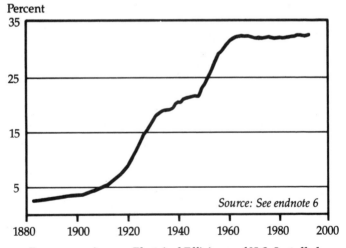

FIGURE 12-1. *Average Electrical Efficiency of U.S. Installed Fossil Fuel Power Plants, 1882–1992*

and Chernobyl. After 1980, nuclear construction dwindled, coming nearly to a halt in many nations by the early nineties.[7]

Electricity demand growth in industrial countries slowed from nearly 8 percent a year in the sixties to an average of 3 percent since the mid-seventies—driven by higher fuel prices and saturation in the use of some appliances, as shown by the levelling off of electricity intensity. (See Figure 12–2.) The confluence of stagnating demand and soaring costs left scores of utilities with multibillion-dollar plants they did not need—creating as of 1994 a $30-billion debt for the state-owned national utility in France and bankrupting U.S. utilities such as the Washington Public Power Supply System and the Public Service Company of New Hampshire. By the early eighties, electricity forecasting had become a

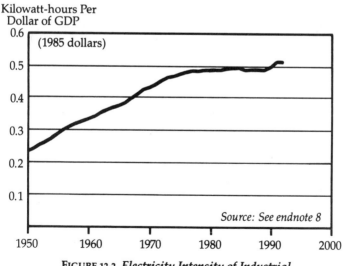

FIGURE 12-2. *Electricity Intensity of Industrial Countries, 1950–92*

guessing game, and utility regulation was transformed from a dull rubber-stamping dominated by the industry into a series of confrontations with consumer advocates and government lawyers.[8]

Further complicating the picture is growing evidence that power plants—particularly the coal-fired models that provide nearly 40 percent of the world's electricity (see Table 12–1)—cause major environmental problems. Increasingly strict pollution laws forced utilities to control everything from coal ash to nitrogen oxide gases. While cleaning up the air, these efforts also drove up the cost of power generation. At the same time, growing concern among indigenous peoples and citizens' groups began to slow construction of numerous power projects, from large hydro dams to nuclear plants.[9]

Managers of electric power systems in developing countries—most of which are government-owned—

TABLE 12-1. *World Electricity Production, 1971 and 1991*

Power Source	1971		1991	
	Use	Share	Use	Share
	(terawatt-hours)	(percent)	(terawatt-hours)	(percent)
Coal	2,142	40	4,671	39
Renewables	1,241	23	2,290	19
Nuclear	111	2	2,106	17
Natural Gas	714	13	1,594	13
Oil	1,102	21	1,376	11
World[1]	5,311	100	12,037	100

[1]Columns may not add to totals due to rounding.

SOURCE: See endnote 9.

were caught in the same wave of problems, aggravated by the fact that they were still in the early stages of electrification. Soaring oil prices and mushrooming debt burdens threw Third World utilities into a period of disarray they have not yet recovered from. Pushed by political leaders to expand supply (as much as 10 percent annually) while cutting prices, utility managers watched their finances and reliability deteriorate. On average, rates of return on investment fell from 9 percent in the early seventies to 5 percent in the eighties.[10]

The World Bank, which plays a leading role in financing electric power projects in developing countries, acknowledged these problems in a 1993 report: "Opaque command and control management of the sector, poorly defined objectives, government interference in daily affairs, and a lack of financial autonomy have affected productive efficiency and institutional performance." In many cases, political corruption made the problem even worse. And given their financial condition, most of these utilities have been unable to add the most basic environmental controls to their power plants.[11]

This mounting series of issues has caused growing turbulence in the electric utilities of many countries, and led to high-stakes struggles over their future. One of the barriers to change is that utilities provide comfortable prosperity for those who control them, and are often used to support powerful political interests. German companies, for example, are required to purchase German coal at several times the world price—and even greater environmental cost—to protect the jobs of coal miners. In Quebec, a costly series of hydro dams is subsidized through the provincial utility, while in France, a massive nuclear industry was created through massive government subsidies. Once such industries are in place—and support thousands of jobs—pulling the plug is no easy political feat.[12]

*   ⋆   ⋆   ⋆*

Efforts to untangle this morass began in earnest in 1978 when the U.S. Congress passed the U.S. Public Utility Regulatory Policies Act, allowing unregulated companies to enter the power market for the first time by building power plants that rely on renewable fuels or that cogenerate heat for industrial facilities. These companies sell electricity to utilities at a price equivalent to the "avoided-cost" of power from utility-owned plants. The new law had its greatest impact in California and Texas, where a competitive breed of independent power producers began to develop everything from large industrial cogeneration projects to wind farms, biomass energy projects, and even solar power plants. More than 6,000 megawatts of renewable generators were installed in California alone, and now provide 11 percent of the state's electricity.[13]

By the late eighties, companies ranging from small entrepreneurial firms to multinational corporations such as Asea Brown Boveri and Texaco were putting their own capital on the line, assuming the risk of cost overruns, and delivering power to U.S. utilities that was often less expensive and more reliable than the average utility-owned plant. The boom in independent power projects coincided with a dramatic drop in utility plant building, so that by the early nineties independent producers were adding about as much capacity as utilities were. (See Figure 12–3.)[14]

Other countries also have begun to open their power grids. In Denmark, Germany, the Netherlands, and

Thousand
Megawatts

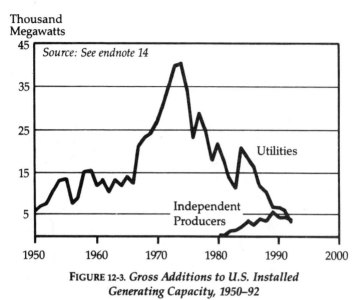

FIGURE 12-3. *Gross Additions to U.S. Installed Generating Capacity, 1950–92*

Switzerland, individual investors have been authorized to put wind turbines or biomass plants on their farms and solar cells on their rooftops. And in the United Kingdom, the government-owned power monopoly was divided in 1989 into 12 distribution companies, an open-access transmission system, and a quasi-competitive generation business. This created a booming market in natural-gas combined-cycle power plants and a smaller boom in wind power, due to a requirement that distribution utilities purchase minimum amounts of "nonfossil" electricity. Recently, China, India, Indonesia, Mexico, and Pakistan also have turned to private companies to build and operate new plants.[15]

Although traditional power plants have been built by independent developers, the emphasis has been on smaller, more innovative power projects. Whereas U.S.

utility-owned generators averaged 600 megawatts per plant in the mid-eighties and 100 megawatts in 1992, plants owned by independent producers averaged just 24 megawatts in 1991. The new generators range from gas turbines to wind turbines and from fluidized-bed coal plants to geothermal generators. Only nuclear power, which has its roots in government and utility monopolies, has been left out of this renaissance of entrepreneurialism in the power industry, although coal has begun to lose favor as well.[16]

Experience in California, Denmark, and elsewhere demonstrates the potential of independent companies to provide cost-effective electricity from relatively innovative and environmentally benign technologies. Many of the new power sources developed by independent producers have been far less expensive than the conventional utility-owned plants built earlier. Today, natural gas and wind are more economical than coal, and even the solar technologies are close to being competitive (even exclusive of environmental costs). (See Table 12–2.)[17]

In the United States, a number of states have set up bidding systems for acquiring new power supplies, ensuring that the most cost-effective projects are selected. The transformation was accelerated by a 1992 law that allows a broader range of companies to enter the independent power market, and that ends utilities' monopoly of interstate transmission lines, permitting wholesale buyers and sellers to use an intervening utility's lines.[18]

Still, independent power generation has a long way to go in many countries. It is outlawed in many parts of the world, and in others it is badly misregulated. These constraints need to be overcome if the new wave of innovation is to continue; fortunately, there are signs of

TABLE 12-2. *Cost of Electric Power Generation in the United States, 1985, 1994, and 2000*

Technology	1985	1994	2000
	(1993 cents per kilowatt-hour)		
Natural Gas	10–13	4–5	3–4
Coal	8–10	5–6	4–5
Wind	10–13	5–7	4–5
Solar Thermal[1]	13–26	8–10	5–6
Nuclear	10–21	10–21	—[2]

[1]With natural gas as backup fuel. [2]No plant ordered since 1978; all orders since 1973 have subsequently been cancelled.

SOURCE: See endnote 17.

change. Growing international communications and trade are encouraging the spread of these ideas, particularly in Europe, where the electric power industry has been one of the last to resist the tearing down of economic barriers between members of the European Union.

★   ★   ★   ★

Even as the power generation market opens to competition, an equally important change is sweeping the electricity industry. Utilities are now spending several billion dollars each year to increase the efficiency with which their customers use electricity, reducing their cost of lighting, cooling, refrigeration, and so on—a new approach to providing electricity services that is known as demand-side management (DSM), or simply "negawatts."

This trend first took root in California in the late seventies when a maverick group of energy analysts reached a startling conclusion: electric utilities and their customers both might be better off if they invested in ways to reduce power use. Studies showed that improving the efficiency with which electricity is used is often less expensive than building and operating new plants. Lawyers from environmental groups brought this argument to regulatory hearings in the United States, and were fiercely resisted by utilities accustomed to promoting power use and building larger and larger generators. Although the environmentalists lost some early battles, several regulatory commissions eventually took their side and ordered the utilities to make improved energy efficiency an active part of their programs.[19]

Early efforts were modest—eliminating discounts for big power users and offering free energy audits to customers, for example. Later, they expanded to include cash rebates for low-energy appliances or compact fluorescent lamps, low-interest loans for home weatherization or industrial retrofits, and even rebates for the purchase of solar water heaters.[20]

Although the notion that it is less expensive to save electricity than to produce it is straightforward, the idea that utilities themselves should discourage the use of their product is counterintuitive. These companies do not operate by the normal market rules, however: their monopolies are protected by governments, and they have access to capital at low interest rates. If all those advantages are applied to building central power stations while capital- and information-short consumers have the full responsibility for efficiency investments, customers end up paying for unnecessary plants that generate power that is used in outdated, inefficient appliances.

The initial approach to remedying this imbalance was to penalize utilities that did not invest in efficiency. This worked for a while, but investments stalled when executives found they were losing money on DSM. The reason: U.S. utilities have traditionally been regulated so that their profits are linked to electricity sales; anything that lowers power use will reduce the return to investors, even if the efficiency investment is added to the capital base on which utility profits are earned. In other words, the economic interests of electricity consumers are at odds with the interests of the shareholders to whom utility executives are responsible.[21]

The solution to this problem was discovered by state utility regulators and advocates who realized, in the words of New England Electric System president John Rowe, that "the rat must smell the cheese." In 1989, the National Association of Regulatory Utility Commissioners proposed that regulators offer "cheese" by allowing utilities to earn equal or greater profits on saved power, compensating utilities in a variety of ways for profits that would be lost from reduced sales. In some cases, companies are now allowed to earn returns on efficiency investments at a higher rate than for building power plants; in others, utilities share directly in the money saved. Such approaches are spreading rapidly in the United States, and also are being tried in the United Kingdom.[22]

Beyond adjusting their incentives, 30 states have ordered U.S. utilities to adopt integrated resource planning. Under this, planners assess the benefits, risks, and costs—sometimes including environmental costs—of all practical electricity generation and savings options. Often the winning option is determined by seeking competitive bids from energy-saving companies as well as power suppliers. Sometimes even an efficiency program that raises power prices may be advantageous, so long as it cuts average consumption and thereby lowers customers' monthly bills.[23]

In response, U.S. utility spending on efficiency grew from less than $900 million in 1989 to an estimated $2.8 billion in 1993. Most of this has been concentrated on the West Coast and the Northeast, where state regulators have been most supportive. Some publicly owned utilities have also moved aggressively to cut power use. A 1993 survey by the Electric Power Research Institute, a utility-funded organization, of most U.S. DSM programs found that they saved electricity at an average cost of just 2.1¢ a kilowatt-hour. Other studies show that poorly managed DSM programs sometimes cost several times as much. Careful design and implementation are therefore essential, including regular monitoring to detect ineffective programs.[24]

Utilities throughout Canada and Western Europe are also adopting demand-side management. Ontario Hydro and BC Hydro, for example, have large programs. More than 50 West European utility-sponsored lighting efficiency programs were launched between 1987 and 1992, and they now exist in 11 nations. Exceptionally strong efforts are found in municipally owned power companies in Denmark, Sweden, and the Netherlands, with Dutch utilities spending around $30 million each year. Although their programs are continuing to grow, most European utility programs still result from simple mandates, and, because they are funded by governments, do not have to be profitable. The giant German utility, RWE, for example, agreed to spend 100 million deutsche marks ($60 million) on DSM over three years as part of a bargain with the government to pursue its first priority—a new coal mine to fuel the utility's power plants.[25]

Developing-country utilities, too, are beginning to pursue improved efficiency. In Thailand, where double-digit power growth is nearly bursting the seams of the energy infrastructure, the national utility has started a $190-million five-year program that includes the purchase of efficient lights, appliances, and motors. Brazil, China, and Mexico are following suit. The Lawrence Berkeley Laboratory in California estimates that efficiency could cut the growth of power use in developing countries by 25 percent over the next 30 years, freeing up billions of dollars. Brazil, for instance, could reduce the projected growth in electricity use in 2010 by 42 percent, estimates Howard Geller of the American Council for an Energy-Efficient Economy. And the transitional economies of Eastern Europe and the former Soviet Union could probably meet their electric needs through 2010 with today's generating capacity just by improving their dismally low levels of efficiency.[26]

Despite the formative stages of many of these programs, U.S. utilities anticipate a decrease in projected electricity use in 2000 of 4 percent. Dutch utilities expect to cut power use projected for the end of the decade by 2.5 percent through lighting programs alone. The most aggressive U.S. utilities, though, are planning to reduce projected sales by more than 6 percent by 2000, with DSM satisfying half their projected growth in power demand. The largest privately owned U.S. power company, Pacific Gas & Electric (PG&E), aims to meet 75 percent of its "growth" via efficiency. As a result of DSM and other programs, California has held per capita power use to the 1979 level through 1992—at the same time that it rose 19 percent in the rest of the United States. (See Figure 12–4.)[27]

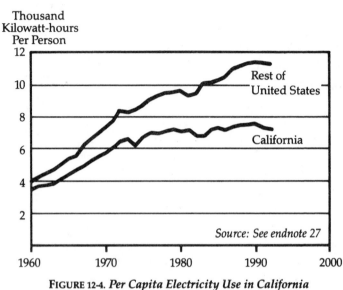

FIGURE 12-4. *Per Capita Electricity Use in California and the Rest of the United States, 1960–92*

If the United States as a whole pursued demand-side management programs as aggressively as California did, growth in U.S. power demand would be cut roughly in half, saving up to $29 billion worth of electricity by 2010. The savings potential in other industrial countries is similar, and in developing countries it is undoubtedly far higher.[28]

*   *   *   *

The shift to investing in improved efficiency in their customers' buildings and factories may also lead the power industry to a new breed of smaller electricity generators. Though the power industry evolved under the assumption that large central stations were the most economical, the advent of more-decentralized power plants relying on gas turbines has already begun to undermine this model. It soon may be obliterated by even smaller, more modular generating and storage technologies that are now rapidly entering the market, including fuel cells, rooftop solar generators, and flywheels. Together, these inventions could make power generation at the household level economical, and at the same time eliminate much of the waste inherent in the current system.

Such technologies have economic advantages ignored by the accounting systems used by most utilities. When a utility sells power to residential consumers, roughly a third of the cost stems from transmitting and distributing the power. (U.S. utilities alone are spending some $11 billion each year on building and upgrading transmission and distribution—one third more than they spend on new power generation.) Allocating these costs more precisely could yield large savings—not only in power plants but in the transformers, switches, and wires that make up the transmission and distribution systems. (The same principle should apply to demand-side management, since it allows a utility to defer investments in power plants as well as transmission and distribution.)[29]

Most utilities recognize only one "avoided-cost" price and do not consider the location of generators or customers. Yet a utility that builds small power plants in a customer's own facility avoids at least some of the distribution cost. In other words, a utility that purchases power for 5¢ per kilowatt-hour from a remote coal-fired power plant should in theory be willing to buy it from a customer's rooftop solar system for up to 8¢ if that power is used locally. The German government in effect recognized this difference in a 1991 law that requires utilities to buy power from household-scale systems at 90 percent of the retail price.[30]

Utility engineers are accustomed to a mix of baseload

plants that operate nearly all the time, with intermediate and peaking plants that are turned on and off to match demand. But the output of solar systems varies according to the weather and time of day, while fuel cells turn on and off with customer demand, requiring new modes of operation. Hourly and daily fluctuations in the availability of wind and sun require utilities to activate backup generators quickly, as well as releasing power from storage systems such as flywheels and compressed air facilities.[31]

Fortuitously, technology is being developed to link each fuel cell, flywheel, and air conditioner electronically, allowing the grid to operate as a single "smart" system that avoids overloading lines and turns off decentralized generators when crews are repairing wires. In an experiment in a subdivision in Little Rock, Arkansas, the Entergy Corporation is installing household-based computer chips linked to a home's television cable. This allows Entergy to limit peak power demand and provide "real time" pricing to its customers, and even lets customers determine when and at what price they buy power. Analysts now believe that distributed, intermittent generators, backed by "peaking" turbines or hydroelectric plants to adjust for minor load swings, can supply up to 30 percent of a typical utility's demand, at *less* than the cost of new centralized generating and transmission equipment.[32]

If deployed properly, distributed generation and storage may increase reliability and reduce costs, as well as reshaping today's utility systems. A utility with 50 generators connected to its system today could see the figure reach 5,000 or even 50,000 by 2010, much as some corporations went from three mainframe computers in 1980 to 30,000 personal computers in 1994. Carl Weinberg, the former research director for PG&E who helped develop the concept of decentralized generation, observes: "Operating modes for utility systems are likely to evolve along a path similar to that taken by computer networks and telephone switching. . . .The networks of future utilities will manage many sources, many consumers, and continuous re-evaluation of delivery priorities. All customers and producers will be able to communicate freely through this system to signal changed priorities and costs."[33]

*   *   *   *

Slowly but unmistakably, the old electric utility model is beginning to crack—and those cracks are beginning to spread. Indeed, in the United States the pressures of increased competition led financial institutions in the mid-nineties to downgrade the bonds and lower the share prices of many electric utilities, which has encouraged them to change even faster. From the ruins of the current industry structure, a broad array of potentially far-reaching experiments is emerging. Still, important questions are being raised about what the electric power industry might look like a few decades from now, and how those changes could affect prospects for a more environmentally sustainable energy system.

A number of industry consultants and even some government regulators now argue that the future lies in a completely open commodity market in electricity. Much of the push in this direction is coming from large industrial users who want to purchase power directly from independent power producers, using the local utility's lines for transmission and distribution—a concept known as "retail wheeling." Although this is not permitted by most governments at the moment, the spread

between the 10–20¢ per kilowatt-hour cost of power
from some utility-owned nuclear plants and the roughly
4¢ per kilowatt-hour cost of wholesale power from new
combined-cycle plants owned by independents has
placed enormous strains on U.S. power systems. Utilities are threatened with the loss of some of their biggest
customers, which would raise prices for others, creating
a potential "death spiral." In the United Kingdom,
where retail wheeling is being introduced, most of the
benefits have gone to moderately large to large industrial
customers.[34]

This threat has led many U.S. utilities to desperate
cost-cutting in order to bring their electricity prices
down. Although there is undoubtedly a lot of fat to be
trimmed, these efforts could easily go too far. The possible adverse consequences include slashed investments
in DSM, distributed generation, and other "energy service" programs that could greatly benefit electricity consumers. Furthermore, a focus on low-cost bulk electricity at the retail level would mainly help large customers
with the most bargaining power—at the expense of
small customers, including not only rich and poor residential consumers but even the small industries that
provide most new jobs. And for utilities themselves, this
could be a risky course. With their cumbersome bureaucracies and expensive portfolios of coal and nuclear
power plants—many of which will cost huge sums to
decommission—few utilities will be able to compete
with independents anytime soon.[35]

Fortunately, another utility model is beginning to
emerge: that of a diversified electricity service company,
committed to providing an array of cost-effective energy
services for its customers. Here, the focus would be not
on achieving the lowest electricity prices but on provid-

ing services at the least cost. (As noted earlier, greater efficiency even with slightly higher prices can mean lower bills.) A strong and well-managed distribution utility would meet the needs of large and small consumers by bargaining for wholesale power and providing investment capital for demand-side management and the construction of decentralized generation and storage devices.

This is the path that the Sacramento Municipal Utility District is following, as are a growing number of other municipal utilities—generally located in forward-looking cities such as Ashland, Oregon, and Saarbrucken, Germany. Many of these smaller utilities have a tradition of purchasing power from larger companies, and think of themselves mainly as distributors, which makes the adaptation to a service model easier.[36]

Similar innovations are being tried by a few investor-owned utilities, though the process is more complex, requiring an overhaul in the way electric utilities are regulated. In the United States, for example, electricity prices are generally determined on a cost-of-supply basis, and "profits" are earned as a percentage of capital investment, which is traditionally concentrated in power plant construction. This system builds in a bias toward selling more electricity—regardless of whether it is being effectively used—and provides little incentive for utilities to improve the efficiency of their own equipment, let alone that of their customers. Most regulators even allow fuel price increases to be automatically passed to customers.[37]

To create a service-focused industry, utilities need to be rewarded not for how much capital they spend but for their ability to provide cost-effective and environmentally clean energy services to their customers—similar to the way U.S. phone companies are regulated. The government will have to set the rules for this new system, but once regulations are in place, there will be less need for intervention, and a less adversarial relationship between utilities and regulators. Under this two-tiered model, distribution utilities would retain a critical "gatekeeping" role, periodically producing integrated resource plans that identify the range of practical investments they can make within the distribution system itself while contracting with wholesale power suppliers for any additional electricity that is needed.[38]

One of the challenges presented under any restructuring scenario is how to ensure that environmental values are appropriately counted in choosing among various power options. Otherwise, short-term economics could lead many regions to become too dependent on natural-gas-fired power plants while neglecting renewable energy sources that have higher capital costs but lower fuel and environmental costs. Traditionally, utilities have only been obligated to meet relevant environmental laws, with no incentive to go any further. But utility regulators in many parts of the world have recently been moving toward mechanisms that ensure that the costs imposed by emissions of various kinds of pollutants are put directly on the table when utilities are considering what kind of a power plant to build or even whether a new one is needed.[39]

Two mechanisms have been proposed to address these issues: environmental costing and set-asides. Massachusetts, Nevada, and New York are among the states experimenting with environmental costing. When new generating options are being evaluated, the cost of pollution is counted, which gives renewables a boost. Meanwhile, California and the United Kingdom have adopted simple set-asides for renewables, in which a minimum share (10–20 percent) of the market for new power is reserved for "nonfossil" generating technologies. Either one of these approaches can work, and some combination of the two may prove particularly effective. A third, supplementary approach being tested by a few utilities, including SMUD, is to sell renewably generated electricity at a slightly higher "green" price to consumers who volunteer, allowing them to subsidize development of technologies such as solar cells.[40]

The advantages of a competitive commodity market in wholesale power, coupled to an efficient energy services market at the retail level, argue for an end to vertically integrated utility monopolies. These two sides of the business are already evolving in opposite directions, exposing an inherent contradiction between a competitive wholesale power market and the "obligation to serve" that most utilities are held to. To achieve the benefits of each, governments would do well to encourage a separation of the industry into wholesale suppliers and local distributors.

The telecommunications industry may provide a model. Though most countries relied until recently on single telephone monopolies, those are now being broken up and local phone systems are being connected to a growing array of long-distance carriers, satellite networks, and cellular systems. In the near future, computer networks and cable television systems will be integrated into this mix, creating a vibrant, competitive industry and melding the once-separate businesses of phone and television into one giant telecommunications market. Although electric power systems will never be as versatile as telecommunications—electricity cannot safely be piped through the air—the power industry's transformation is likely to have broad similarities, driven in part by the same digital technologies.

If it follows the telecommunications model, the electric power industry will consist of a competitive commodity market in wholesale electricity and a service-focused distribution business connected through a "common carrier" transmission system that could either be owned independently or be jointly owned by the distribution utilities. Most existing power plants would at some point be sold to independent power producers, while many high-cost nuclear plants were written off as tax losses, with regulators deciding how their excessive costs—now estimated at $25–30 billion a year in the United States alone—are to be shared between the utility and its consumers. Disposing of those "stranded assets" will keep many lawyers and bankers busy for years.[41]

Still, the biggest growth markets in the next decade may turn out to be on the distribution side of the business—investing in everything from refrigerators to solar cells. Indeed, this trend may point many utilities into the telecommunications business itself. Steve Rivkin, a Washington, D.C.–based lawyer, has suggested that power companies could find it cost-effective to make an investment in the "information superhighway," linking their customers with a cable that provides everything from electricity load management to cable television and computer services. This may lead to some creative partnerships between electric utilities and phone and cable companies.[42]

Of course, no single model will work equally well everywhere, and new approaches will have to be adapted to local conditions. Developing countries face particular challenges in providing electricity services for rapidly growing economies. So far, many have begun purchasing power from private generators, but few have developed an institutional framework for improving energy efficiency and decentralizing the power system. This is unfortunate, because this new model may have even greater benefits in countries that are faced with the need to invest large sums in their power systems in the years ahead. The move to a competitive, service-focused business can encourage increased efficiency in power supply as well as in energy services, reducing the amount of capital required. Moreover, decentralized generators may provide more-reliable service in areas now plagued by power outages.

For the world as a whole, the shift to a competitive market for wholesale power linked to a service-oriented distribution system offers both economic and environmental advantages. Indeed, we can foresee a time several decades from now when the electric power industry might produce minimal pollution, while at the same time serving as a catalyst for the development of a range of exciting new automotive technologies and building designs.

The electric power industry was formed to meet the challenge of providing power for millions of people who did not have it. Although it has met that challenge well—at least in industrial countries—the industry is now faced with an equally profound environmental challenge. Both can be met, but not without major changes in the industry's structure. As David Freeman says, "We're at a point in our history where the electric utility industry can start wearing a white hat. We can be the organization that brings cleaner air to our children and economic vitality to our cities."[43]

Meanwhile, Freeman has taken his once-radical ideas to yet another utility in need of a shake-up. In early 1994, he was appointed by Governor Mario Cuomo to head the New York Power Authority.[44]

IV

Energy Futures

13

Through the Looking Glass

In 1968, a team of analysts in the London offices of the Royal Dutch Shell company was at work on a study of the global oil market. As one of the world's largest petroleum companies, Shell had an obvious interest in anticipating future developments. But like most of the top people at other petroleum firms, Shell's executives comfortably assumed that the future would be much like the recent past—marked by steady or declining prices, and growing demand. Yet the Shell planning team, headed by Frenchman Pierre Wack, began to spot signs of trouble in the late sixties. World oil consumption was skyrocketing, and production in the United States was levelling off. Middle Eastern nations were rapidly increasing their share of the world oil market and beginning to show signs of restiveness about the low oil prices forced on them by Shell and other major oil companies.[1]

Wack and his colleagues developed two separate projections of future oil trends. The first was a low-oil-price, "business-as-usual" forecast; the second described a future crisis in which the Organization of Petroleum-Exporting Countries (OPEC) gained ascendancy and oil prices skyrocketed. When his bosses did not react to the new information, Wack tried another approach. Instead of simple numerical projections, he developed full-blown "scenarios"—stories about the future that described the reasons for and the full ramifications of a serious oil crisis, including the possibility that major companies' dominance of refining and marketing might be threatened by OPEC. This approach had a far greater impact on Shell executives, who began debating the company's future. Although its strategic plans were not immediately overhauled, when the first oil price hike struck in late 1973, Shell was far better prepared than most other oil companies. Within a few years, it had become the world's most profitable petroleum firm.[2]

The techniques developed at Royal Dutch Shell have since gained a following among business and government planners, thanks to Peter Schwartz, who worked there briefly in the seventies and is now president of the Global Business Network, a California firm providing strategic advice to companies. Schwartz points out that scenarios—derived from the theatrical word for script—are not intended to predict the future, a task better left to fortune-tellers. "Scenarios are stories about the way the world might turn out tomorrow, stories that can help us recognize and adapt to changing aspects of our present environment," Schwartz says. "The purpose of scenarios is to change your view of reality—to match it up more closely with reality as it is, and reality as it is going to be."[3]

Despite the success of Schwartz and his colleagues at marketing the scenario planning concept to dozens of companies, its impact in the energy sector has been negligible. Indeed, most energy companies and government agencies are still considering the future much the way that Shell executives did in 1968. Increasingly powerful computers are used to churn out ever more detailed projections of future trends, but most consider just a narrow band of macroeconomic assumptions, often limited to a few different oil price levels and rates of economic growth. They generally assume limited if any technological progress, and major discontinuities or surprises—particularly unwelcome ones—are rarely considered.

This narrowing of options is often exacerbated by the tendency to a sort of corporate herd mentality. Energy consultant Daniel Yergin noticed some years ago that oil price forecasts tend to cluster around the same figure—and then suddenly jump as a group to another level. When a forecaster was asked about this tendency, he replied candidly that when his bosses asked him to justify his projection, it helped if he could point out that he was halfway between Exxon and Shell. This tendency is reinforced by the state of denial that plagues many energy corporations. Oil and coal companies do not want to consider that their products may face a long period of decline, while the automobile and electric power industries are reluctant to face the massive new investments and restructuring that a change of course might suggest.[4]

The World Energy Council (WEC), an organization that includes most of the world's largest private and government energy institutions, exemplifies the problems.

Every three years, WEC holds a widely publicized World Energy Congress, and prepares an assessment of future energy trends. Its latest effort, published in 1993 after a two-year research effort involving 500 experts and nine regional teams, foresees an increase in world oil use of some 60 percent by the year 2020 and a doubling in the use of coal. The Council's baseline scenario—the one it says is the most likely—would continue the trend toward heavy use of electricity provided by central power plants. While use of natural gas is expected to increase, according to WEC, so is reliance on the most polluting fossil fuels, including tar sands and oil shale. In sum, the World Energy Council envisions the basic outlines of today's energy systems remaining largely unchanged for decades to come.[5]

The WEC study does consider other scenarios, but they generally cover a narrow range of assumptions. Even the relatively innovative "ecologically driven" path that is included in the 1993 version is tightly constrained. Here, higher levels of efficiency would allow somewhat lower levels of fossil fuel use—and consequent reduction in emissions of carbon dioxide and other pollutants. But the study goes on to question the feasibility of this scenario. Indeed, Council representatives have stated that their study proves the impossibility of meeting the central goals of the Rio climate treaty, which include holding carbon emissions in the industrial countries to 1990 levels and, later, stabilizing atmospheric concentrations of carbon dioxide.[6]

This mischaracterization by the World Energy Council of its own results is symptomatic of the problems with most such studies. Although the assumptions used have evolved somewhat since the mid-seventies—projections made then now appear ridiculous—they continue to lag behind changes in the real world. Among the blinders that skew all the WEC scenarios are extensive analysis of fossil fuel reserves but no quantitative estimation of renewable resources; the lack of a detailed assessment of energy-using appliances and industrial equipment; extensive discussion of the pros and cons of nuclear power, but hardly a mention of solar photovoltaics or fuel cells; and the assumptions that technological advancement will proceed slower in the future and that developing countries will continue to lag far behind the industrial world in both economic and environmental development.

Many of these assumptions seem dubious; others are frankly implausible. Overall, they paint a misleadingly gloomy picture of the potential to forge a more sustainable energy economy in the decades ahead. Sadly, they are the same assumptions used by most governments and by organizations such as the International Energy Agency.

* * * *

New thinking often comes from unexpected quarters. Since shortly after Charles Darwin wrote *On The Origin of Species*, biologists have tended to think of evolution as an exceedingly gradual, step-by-step process, with an almost infinite number of incremental stages between one species and the next. During the seventies, however, Harvard biologist Stephen Jay Gould developed an alternative theory known as "punctuated equilibrium," which posits that most evolutionary change occurs in sudden bursts—driven in part by external environmental influences that force species to evolve quickly in order to survive. According to Gould, evolution is often marked by long periods of stasis, followed by brief eruptions of radical change, a theory that has since garnered broad acceptance among biologists.[7]

Technology, oddly, may follow a similar path. Take the telephone, for example, which developed rapidly in the late nineteenth century and then became virtually static in the middle decades of the twentieth century. Now the telephone is again in a period of rapid transition—simultaneously becoming digital, wireless, and portable, while also being connected to fax machines, computer modems, and so on.

Punctuated equilibrium has a similar relevance to energy. Indeed, as described in Chapter 2, energy technologies underwent a period of massive change in the last decade of the nineteenth century and the first decade of this one. The modern energy economy was created virtually out of whole cloth during that period; by 1910, many U.S. cities had been transformed: on the streets, horses had been replaced by automobiles, while candles and gas lights had been supplanted by electric lights. Such transitions are usually driven by an array of social and economic forces, coupled with the availability of technologies that can be applied in new ways.

If these are the conditions needed for rapid change, then the last decade of this century and the first decade of the next may be as revolutionary as those of 100 years ago. In Part I we discussed the growing array of economic and environmental forces of change that are beginning to shape the world energy economy. Environmental conditions are changing rapidly in many regions, and public attitudes are shifting with them; energy institutions are being profoundly restructured around the world, and key technologies are progressing at an explosive pace. In many respects, the world energy economy is already in the midst of a period of change that is more dynamic than anything we have seen for at least eight decades.

Still, change takes time, and an array of impediments continue to block the adoption of new, more efficient technologies and practices. The Rio climate treaty includes no binding limits on carbon emissions, few countries have adapted their energy taxes so as to take account of the climate issue, and a disturbingly large share of government energy R&D funding is going to programs with virtually no chance of producing anything useful. To jump from the business-as-usual path that most governments and corporations are planning for to the quite different path that we think is possible, additional "stress" on the world's economic and political systems will likely be required. To consider how that might happen, we need to develop our own story about the future.[8]

The date is early September 1998, and many parts of the world have just experienced one of the most brutal summers in memory. A severe drought and heat wave have cut the North American grain harvest to 20 percent below normal, following a 10-percent shortfall the year before. In an unusual coincidence, droughts have also reduced grain harvests in Australia and China, while a series of fierce summer rain storms damaged crops in Europe. A decade earlier, such disasters might have caused only a minor ripple in commodity markets, but the excess grain reserves of the eighties are a distant memory. The world went into the 1998 growing season with storage bins nearly empty. Rapidly growing food demand— particularly in Asia—is outstripping supplies. By mid-September, the prices of corn and wheat have tripled, and a consumer revolt forces the U.S. government to restrict grain exports and halt food aid. Rising grain prices cause a dra-

matic worsening of the famines that first appeared in parts of
Africa and South Asia in the mid-nineties.

Then in late September, with newspapers still full of news
of the food crisis, a Class 5 hurricane steams out of the Caribbean and slams into New York City. A week later, a similar
storm strikes New Orleans. Each is larger than Andrew, the
devastating storm that hit south Florida in 1992, and because the 1998 storms hit major urban centers, the economic
damage is far greater. Early estimates place losses at $50
billion, sufficient to send the stock prices of several insurance
companies reeling. A few weeks later, a cyclone strikes south
of Dacca, the capital of Bangladesh. The storm is similar in
magnitude to the ones in the United States, but rapid growth
in the number of people living along the mouth of the Ganges
has left millions vulnerable. Although the government does
not announce a casualty figure, U.N. relief officials estimate
that as many as a million Bangladeshis have died.

In mid-October, one of the most dramatic scientific press
conferences ever held takes place in a Geneva ballroom.
Called by the Intergovernmental Panel on Climate Change
(IPCC) and attended by 130 leading climate scientists, the
message is simple and disturbing: advances in the understanding of atmospheric dynamics show that rising concentrations of carbon dioxide have increased the frequency and
severity of catastrophic droughts and storms. The scientists
warn that the situation could grow far worse in the next two
decades, and urge evacuation of many low-lying areas. Carried live on CNN, the press conference causes shock waves in
many countries. Headlines the next day proclaim: "Unnatu-
ral Disasters Threaten Millions." The world's stock ex-
changes, still reeling from the calamities of the previous
month, plunge again.

In early November, a special session of the U.N. General
Assembly is convened to consider the climate threat. In an
unprecedented joint appearance, the President of the United

States and the newly elected 42-year-old Prime Minister of China address the assembled ministers. Representing the world's largest carbon emitters, the two leaders call for quick action to reduce the use of fossil fuels and slow the pace of climate change. Each announces unilateral actions to cut carbon emissions. The Prime Minister says that his government has concluded that coal-fired air pollution is causing so much damage within China that, for purely domestic reasons, China will be better off following a different path. The U.S. President observes that the potential market for technologies such as fuel cells and photovoltaics is so large that the government wants to get a head start in "growing these new industries and creating good American jobs."

The next few months are extraordinarily hectic. Round-the-clock climate negotiations commence in Geneva, and within weeks, binding carbon emission limits have been adopted by international climate negotiators, while 17 countries have announced plans for sizable carbon dioxide taxes. Burning coal is becoming as popular as cigarette smoking was in the early nineties. Canada and Germany announce plans to end coal combustion within a decade, and India and Russia say they will cut coal's use by 50 percent.

In December, the grim news is interrupted by a series of announcements suggesting that the renewable energy business is taking off. General Electric reveals plans to invest billions of dollars in wind power manufacturing, while Siemens and Mitsubishi launch even larger efforts to expand their commitment to photovoltaic and solar thermal energy development. And in a step reminiscent of the media megamergers of the early nineties, Amoco and Asea Brown Boveri declare that they are joining forces to build an array of hydrogen pipelines and fueling stations that will stretch from Los Angeles to Washington, D.C. In early January 1999, a surprise press conference is called at Toyota City in Japan. The world's eight leading automakers disclose plans to accelerate intro-

*duction of a new generation of hybrid-electric cars that they
have been working on for several years, then go on to call for
a phaseout of gasoline-fueled internal combustion engines by
2010.*

Although the chances of this scenario being played
out in just the terms described are practically nil, its
general outlines are broadly plausible. Indeed, the likelihood that future energy trends will be smooth and predictable, presenting decision makers with no unexpected crises, seems far more remote. And as General
Motors and IBM have already discovered, planning for
a static future can be dangerous to the health of companies. Now is the time for corporate and government
leaders to consider carefully what they might do if outside forces push world energy trends in a sharply different direction.

To consider this sort of "sustainable future" more
fully, we have charted the kind of energy scenario that
might emerge if the need for change becomes more urgent. More specifically, we ask what sort of energy path
might unfold if the world decides to take the overriding
goal of the Rio climate treaty seriously—stabilizing atmospheric concentrations of carbon dioxide by the middle of the twenty-first century. The "business-as-usual"
scenario used by IPCC shows carbon dioxide concentrations rising from the 1993 level of 357 parts per million to about 700 parts per million at the end of the next
century, while the World Energy Council has it reaching
about 600. (See Figure 13–1.) Our scenario, on the
other hand, would have carbon dioxide concentrations
levelling off by mid-century at about 450 parts per million, and then declining gradually in the following
decades.[9]

This scenario is based on the assumption that govern-

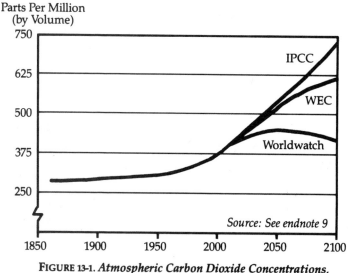

FIGURE 13-1. *Atmospheric Carbon Dioxide Concentrations, 1860–1993, With Scenario Projections to 2100*

ments would rationally but aggressively pursue the technologies and policies discussed in earlier chapters. No breakthroughs in cold fusion or government takeovers of the energy sector are considered. Indeed, we believe that our "sustainable future" will unfold most rapidly in market-based economies, although governments will need to reverse many of today's energy subsidies. (The policies needed to achieve this are described in some detail in the final chapter.)

In short, ours is an energy path characterized by high levels of efficiency, extensive use of decentralized technologies, heavy reliance on natural gas and on hydrogen as an energy carrier, and a gradual shift to renewable energy sources. The outlines of this scenario and a comparison with those of the World Energy Council and other organizations can be seen in Figures 13–2 and 13–3.[10]

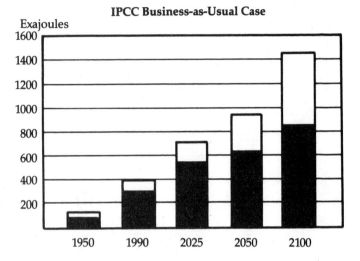

IPCC Business-as-Usual Case

Exajoules

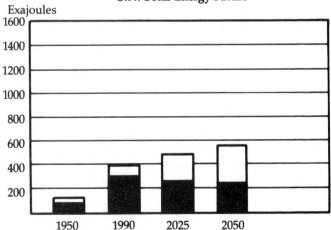

U.N. Solar Energy Future

Exajoules

FIGURE 13-2. *World Primary Energy Use by Scenario, 1950–2100*

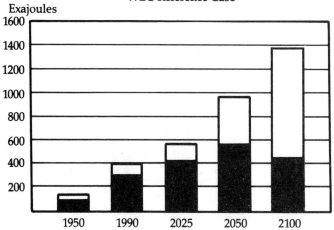

WEC Reference Case

Exajoules

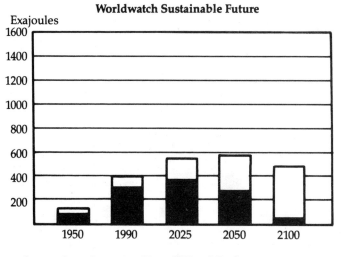

Worldwatch Sustainable Future

Exajoules

Source: See endnote 10   Key: ■ Fossil Fuels □ Other Sources

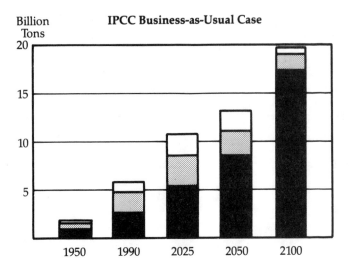

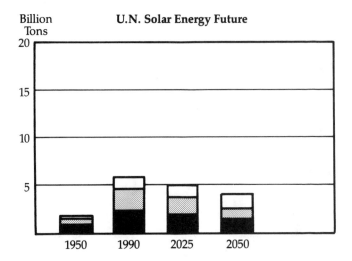

FIGURE 13-3. *World Carbon Emissions by Scenario, 1950–2100*

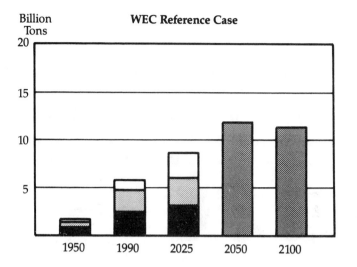

Billion
Tons
WEC Reference Case

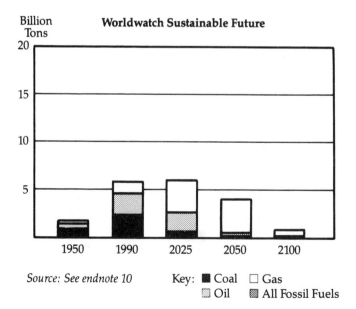

Billion
Tons
Worldwatch Sustainable Future

Source: See endnote 10   Key: ■ Coal ☐ Gas
                                      ▨ Oil   ▨ All Fossil Fuels

In constructing this picture of a sustainable future we have assumed that global economic growth averages 2.5 percent per year, similar to the figures used in most studies. In contrast to them, however, we have assumed a slowing of population growth along the lines called for at the 1994 International Conference on Population and Development in Cairo. Higher estimates are inconsistent with the carrying capacity of many regions—particularly Africa—and even at the slower rate assumed here, world population would reach nearly 8 billion before peaking at mid-century. On this path, 85 percent of the population in 2050 will live in what are now called developing countries, up from 78 percent in 1990.[11]

Virtually all analysts agree that efficient use of energy is the cornerstone of a more sustainable energy system. In order to hold carbon emissions to about the current level in 2025, and then cut them substantially, we estimate that the world will need to double the current level of global energy productivity—the amount of energy needed to produce a dollar of gross world product—over the next four to five decades. By comparison, it took the United States 72 years to double the level of energy productivity it had in 1921.[12]

This goal will be made more challenging by the fact that developing countries now have average rates of energy use just one fifth the current level in Europe, and are rapidly building fossil-fuel-powered factories, buildings, and transportation systems. Still, our projected increase in energy productivity is consistent with the potential gains identified in several other studies. Indeed, the goal seems less impossible when a comparison is drawn with the tenfold increase in U.S. labor productivity during the past century.[13]

The second key to stabilizing carbon dioxide concentrations in the atmosphere is a shift in the mix of fossil fuels. Our sustainable future scenario posits a 73-percent reduction in coal use by 2025, and a 20-percent cut in oil use, some of which would be replaced by natural gas. Gas is less carbon-intensive than the other fossil fuels, and lends itself to efficient applications, including potentially widespread cogeneration of electricity and heat in factories and buildings. As argued in Chapter 5, natural gas resources appear adequate to permit a tripling in global production by 2025. Although such estimates are somewhat speculative, our relatively conservative figures suggest that world gas consumption would peak by about 2030, fall sharply after 2050, and be largely phased out by the end of the next century.

The third step to a sustainable future is the development of carbon-free energy sources. Although past studies assumed that nuclear fission or fusion would fill this niche (and the World Energy Council still makes that assumption), that hypothesis is no longer credible. The current trend in most countries is away from nuclear power; continuing concern about nuclear waste and safety and the recently uncompetitive cost of nuclear power suggest that this trend will continue. We assume a gradual decline in nuclear power between 2000 and 2025 as aging plants are retired.[14]

As fission fades, nuclear fusion seems increasingly unlikely to ride to the rescue. Scientists working in government-funded fusion programs in Europe and the United States do not foresee it being introduced on a commercial scale before 2030, while skeptics argue that the long-term prospects for fusion are dubious as well. Even if current development efforts succeed, the large power plants that would likely result run counter to the trend to decentralized, modular technologies that is now under way.[15]

As noted in earlier chapters, the prospects for renewable sources of energy are far brighter. Currently only biomass and hydropower are used in sufficient quantity to show up in world energy statistics, accounting respectively for 13 percent and 6 percent of world primary energy use. These shares would grow in the next few decades in our sustainable future, but in both cases, growth would be constrained by resource limitations. At the same time, the so-called new renewables—solar, wind, and geothermal technologies—that play a minimal role today are likely to take off within the decade, as costs decline and production is scaled up.[16]

Under our scenario, wind power, solar photovoltaics, and solar thermal energy would each provide about as much primary energy in 2025 as nuclear power does today. Each would be used primarily for electricity generation, though in different ways. Solar thermal and wind power stations—roughly 1,500 gigawatts each—would be deployed across resource-rich areas such as the U.S. Great Plains, the Thar desert in India, la Ventosa in Mexico, and the Sahara in north Africa. Solar photovoltaics would be more widely distributed, mainly on the roofs and facades of buildings.[17]

In order to continue the trend away from fossil fuels after 2025, a 75-percent increase in the harnessing of renewable energy will be needed between 2025 and 2050. By then, renewables would displace oil as the world's second largest energy source, providing over half the world's primary energy, with the share rising as high as 90 percent by 2100.

Although an energy system this different is hard for most people to envision, we see no technical or economic barriers to such a transition. The projected annual growth in new renewable energy technologies—as high as 20–30 percent—is actually slower than the growth rates of nuclear power in the sixties and seventies, or of personal computers in the eighties. Under this scenario, the renewable energy industry would have annual revenues of roughly $200 billion (1993 dollars) in 2025—twice the 1993 revenues of Exxon.[18]

*   *   *   *

Among the aspects of a new global energy system that are most difficult to picture is how—especially in the absence of liquid petroleum—energy would be stored, transported, and distributed to customers. Most recent efforts to anticipate the energy system of the next century assumed a big increase in the role of electricity and a continuing dependence on liquid fuels for transportation. But electric power is an expensive form of energy that is difficult to store, which is one reason that it only meets about one eighth of total end-use energy needs today. Relying on electricity as the world's main energy carrier would require using it for many inappropriate applications such as heating and cooling, which account for over half of current energy use in industrial countries.[19]

Other experts believe that we should try to preserve the current liquid-fuels infrastructure by liquefying coal or biomass—turning them into methanol or another fuel. Yet methanol is a poisonous fuel that readily mixes with water and causes health problems if it comes in contact with the skin. Moreover, methanol from carbon-intensive coal would exacerbate rather than diminish the greenhouse effect, while methanol from biomass would create other problems. Given the stresses already affecting many biological systems and the scarcity of cropland and irrigation water that are projected for the near future, heavy reliance on methanol from biomass would likely be disruptive and unreliable. Such scenarios also assume that methanol would be carried in large quantities from Africa and Latin America, where it would be produced, to the large energy markets of Europe and East Asia—in effect recreating many of the problems with today's petroleum economy.[20]

Another, more appealing vision beckons. Humanity first relied on solid fuels—moving from wood to coal—and then began to shift to liquid oil early in this century. In recent decades, a new trend has developed, as natural gas has begun to displace both liquid and solid fuels in many applications. This shift to gas is in effect a continuation of a long-term trend to ever more efficient, less carbon-intensive fuels—part of the gradual "decarbonization" of the world energy system. Figures for the United States show the broader trends, though recent patterns are distorted somewhat by the disruption in gas markets that occurred during the seventies. (See Figure 13–4.) As natural gas enters the transportation market in the next decade, there will be no major segment of the energy economy from which it is excluded.[21]

As natural gas supplies level off or are voluntarily kept in the ground in order to reduce carbon emissions, a substitute will be needed in the decades ahead. The fuel most likely to fill this niche is hydrogen, the simplest of the chemical fuels—in essence a hydrocarbon without the carbon. Hydrogen is the lightest of the elements as well as the most abundant. When it is combined with oxygen to produce heat or electricity, the main emission product is water.[22]

The logic of a transition to hydrogen has been argued by scientists for more than a century. In the 1870s, Jules Verne wrote that hydrogen would be a good substitute for coal. Although hydrogen has a reputation as a partic-

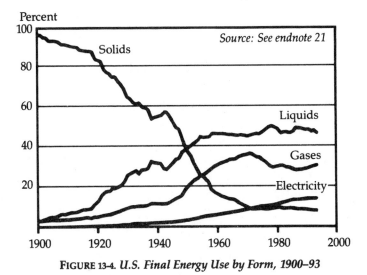

FIGURE 13-4. *U.S. Final Energy Use by Form, 1900–93*

ularly dangerous fuel, this is largely a myth. It can be explosive in the right conditions, but this is true of gasoline and natural gas as well. If properly handled, hydrogen will probably be safer than the major fuels in use today.[23]

Small amounts of free hydrogen are contained in natural gas, but it is generally not found in that form in the natural environment. Rather, hydrogen is bonded with other atoms in more complex molecules that can be broken apart to extract the hydrogen. The most abundant source is water. Electricity can be used to split water molecules through electrolysis, a century-old technology already used commercially. Although it is relatively expensive today, costs would come down as the technology is scaled up. As far as water is concerned, the requirements are relatively modest. In fact, all current U.S. energy needs could be met with just 1 percent of today's U.S. water supply. Even in most arid regions, water requirements will not be a major constraint on hydrogen production. The water needed by a photovoltaic power plant producing hydrogen is equivalent to just 2.7 centimeters of rain falling annually over an area the size of the plant. And in the long run, hydrogen may be derived from seawater.[24]

The real challenge comes in finding inexpensive sources of electricity that can be used to split water. Most of the early hydrogen advocates came from the aerospace and nuclear industries, and they developed a centralized vision of a hydrogen-powered economy. Some predicted massive offshore nuclear islands that would produce enough hydrogen to serve whole countries; others suggested that orbiting satellites could beam concentrated solar energy to gargantuan hydrogen production centers on earth. None of these schemes is remotely practical or economical, and all ignore one of the key advantages of hydrogen: the equipment to produce it is almost as economical on a small scale as on a large one. Both in production and use, hydrogen lends itself to a decentralized system in which waste is minimized.[25]

In our scenario, the obvious candidates for hydrogen production are wind and solar energy, supplemented by biomass gasification. As large wind farms and solar ranches appear in sunny and windy reaches of the world, they can generate electricity that is fed into the grid when power demand is high, and produce hydrogen when it is not. Additional hydrogen could be produced in individual homes and commercial buildings using rooftop solar cells. Hydrogen can either be stored in a basement tank for later use or be piped into a local hydrogen distribution system. In either case, it would gradually fill the niches occupied by oil and natural gas today—including home and water heating, cooking, industrial heat, and transportation. Hydrogen-powered cars have already been developed by several companies, with the main future challenges being an improved storage tank and an inexpensive fuel cell engine.[26]

The key to the practical use of hydrogen is efficiency. According to research carried out at Princeton University and the German Aerospace Research Establishment, even if the marginal cost of the electricity used to produce hydrogen were only 1–2¢ per kilowatt-hour, which may be possible in well-sited facilities built to produce power as well as hydrogen, the delivered cost of the fuel would still be more than three times as much as U.S. consumers now pay for natural gas, though roughly equal to the price that European consumers now pay for gasoline (including taxes). But hydrogen lends itself to highly efficient applications, and as a result, the actual delivered costs of energy services to customers could well be lower than they are today.[27]

One of the keys to extremely efficient use of hydrogen is the fuel cell, which can produce electricity directly from the fuel at an efficiency as high as 65 percent. Indeed, the fuel cell may one day be thought of as the silicon chip of the hydrogen economy. Many homes could have reversible fuel cells—capable of producing hydrogen from electricity and vice versa.[28]

The transition to a hydrogen energy system raises other important questions. We currently depend on large quantities of petrochemicals—used in everything from fertilizers to cleaning fluids and clothing. These carbon-based materials are derived from the various oil by-products that flow in enormous quantity from the world's refineries. If oil is abandoned, we will need another source. Two options beckon: in the short run, natural gas can be separated into hydrogen and carbon—with the latter going into useful materials rather than the atmosphere; in the long run, we can produce petrochemicals from crops on a sustainable basis, which is a far more important use of cropland than simply having it produce fuels—and will require less land.

Eventually, much of the world's hydrogen is likely to be carried to where it is needed through pipelines similar to those now used to carry natural gas. This is more efficient than the oil or electricity distribution systems in place today. Power lines, for example, are expensive to build, and over a distance of 1,000 kilometers lose 3–5 percent of the electricity sent. Moving energy in the form of a compressed gas is substantially less expensive, though hydrogen will cost 30 percent more than natural gas to move, just because it is a lighter element that is more difficult to compress. In the early stages of the transition to hydrogen, the new energy gas can be added to natural gas pipelines in concentrations up to 15 percent—a clean-burning mixture known as hythane. In the long run, engineers believe that it will not be too difficult to modify today's natural gas pipelines so that they will be able to transport hydrogen. Hydrogen could also be produced from natural gas, either in central facilities or right at the gas station.[29]

Over time, solar- and wind-derived hydrogen could become the foundation of a new global energy economy. All major population centers are within reach of sunny and wind-rich areas. The Great Plains of North America, for instance, could supply much of Canada and the United States with both electricity and hydrogen fuel. The pipelines that now link the gas fields of Texas and Oklahoma with the Midwest and Northeast could carry hydrogen to these industrial regions. Although renewable energy sources are more abundant in some areas than others, they are far less concentrated than oil, with two thirds of proven petroleum reserves being in the Persian Gulf.[30]

For Europe, solar power plants could be built in southern Spain or North Africa. From the latter, hydrogen could be transported into Europe along existing gas pipeline routes. To the east, Kazakhstan and other semi-arid Asian republics could supply energy to Russia and central Europe. For India, the sun-drenched Thar Desert is within easy range of the rest of the country. For the more than 1 billion people of China, hydrogen could be produced in the country's vast central and northwestern regions and shipped to the population centers on the coastal plain. And in South America, the wind resources of Patagonia could become a valuable energy resource for the entire southern part of the continent.

Many people assume that producing sufficient hydrogen and other fuels from renewable energy sources requires such huge swaths of land that extensive dependence on them is impractical. In fact, solar and wind energy are far less land-intensive than many of the energy sources now in use. (See Table 13–1.) Today's giant hydro dams and coal strip mines claim extensive land areas—often rendering them unusable for anything else for centuries.[31]

The amount of land required for renewable energy development under our scenario is surprisingly modest. To provide the world with the 55 exajoules of solar thermal energy needed in 2050, an area about the size of Costa Rica or Bhutan, or less than a sixth the size of Arizona, is needed. To supply the 50 exajoules of wind energy called for in our scenario, wind farms would

TABLE 13-1. *Land Requirements for Electric Generation Technologies*

Technology	Land Requirement
	(square kilometer-years per exajoule)[1]
Dedicated Biomass Plantation	125,000–250,000
Large Hydro	8,300–250,000
Small Hydro	170–17,000
Wind[2]	300–17,000
Photovoltaic Central Station	1,700–3,300
Solar Thermal Trough	700–3,000
Bituminous Coal	670–3,300
Lignite Coal	6,700
Natural Gas-Fired Turbine	200–670

[1]Averaged over assumed 30-year life cycles for power plants, mines, and so on. [2]The lower range for wind includes only land occupied by turbines and service roads, while the higher number includes total area for a project.

SOURCE: See endnote 31.

spread over an area about the size of Viet Nam, or less than the area of Montana. (Only one twentieth of the area covered by a wind farm would actually be occupied by turbine towers and service roads, however, leaving the rest of the land for crops or livestock.) The land needed for solar and wind development is in each case small enough that environmentally sensitive areas can be withheld from development without significantly diminishing the energy that can be harnessed.[32]

★   ★   ★   ★

Although the availability of sufficient land, fuels, and technologies to support a sustainable future can be demonstrated, important questions remain about the respective roles of rich and poor countries during the transition. Developing countries, with more than three quarters of the world's population, account for just one fifth of the carbon dioxide released to the atmosphere during the last century, a figure that has risen only to 33 percent in recent years. Although per capita emissions in industrial countries still average 10 times those in poor nations, as the latter nations industrialize, they become central to any effort to stabilize the climate.[33]

In the past, many experts assumed that developing-country energy use (and carbon emissions) would gradually rise toward the levels in industrial countries. But if the world is to hold emissions to the 1990 level in 2025 and then gradually reduce them, another path is needed. Under our scenario, developing-country emissions would increase, but the more profound trend would be a steep decline in emissions in industrial countries—by half by 2025. The net result is that per capita emissions would tend to converge at roughly one fifth the current level in industrial countries.[34]

Although many Third World energy officials fear that carbon dioxide limits could choke off their economic growth, the greater risk would come in failing to use the more efficient, less polluting energy technologies that industrial nations are beginning to adopt. Continuing to pursue an inefficient oil- and coal-based future would saddle developing countries with a grim combination of uncompetitive technologies and economically draining environmental cleanup bills. Indeed, Russia has already tried this path to economic development.

Still, there is reason for optimism. When countries industrialize later—as Germany and Japan did in the fifties and sixties—they generally leapfrog to a higher, more-efficient level of technology. Thus it is possible that by 2025, countries such as Hungary or South Korea could be more energy-efficient than Japan. And it would not be surprising to find that by mid-century China has a more ambitious solar-hydrogen system than Europe does.

How quickly might the transition unfold? When oil prices first soared in the seventies, energy markets responded slowly at first, but then quickened. Government responses were initially misguided, but gradually the more foolish projects were abandoned and better policies emerged. The reaction to a serious climate crisis might well be quicker. The world has been laying the policy and technical groundwork for a new energy system for two decades; with sufficient political pressure, it could accelerate the process dramatically. In Chapter 14, we discuss the policies that could make a difference.

Scenarios are of course not predictions, and the future will inevitably be far more complex than portrayed here. If the past is any guide, unexpected events, new scientific developments, and technologies not yet on the drawing board could push the pace of change in unexpected directions. Still, we are reasonably confident that the vision we have painted here is conservative, and that change may unfold more rapidly than even optimists dare consider.

14

Launching the Revolution

When the leaders of 106 nations met at the Earth Summit in steamy Rio de Janeiro in June 1992, few doubted the historic nature of the gathering. But after all the inspiring speeches and media hoopla, it was the two global treaties agreed to in Rio that were its signal accomplishments. One of these, the Framework Convention on Climate Change, commits the world community to an extraordinary goal: bringing stability to the global atmosphere on which we all depend.[1]

During the two years following Rio, as the treaty slowly reached and then surpassed the 50 ratifications needed to bring it into force—Portugal was the fiftieth—the significance began to sink in. The world has now pledged to fundamentally transform an energy system that has served us well for many decades but that now puts our future at risk.[2]

Sadly, most of the leaders gathered at the Earth Summit went home to domestic energy policy frameworks that are a far cry from the lofty rhetoric of Rio. The list of countries that ratified the treaty in the first year hints at the profound challenge ahead—ranging from the desperately poor country of Mauritius, which still relies heavily on wood fuel, to the United States, the world's largest fossil fuel user (and one of the most reluctant treaty signers), to the tiny Maldives, an island nation whose very existence may be threatened by climate change. In one country after another, heavy subsidies for coal, research budgets dominated by unpromising nuclear technologies, and uncompetitive market structures are among the extensive barriers to reforming the world energy system.[3]

As earlier chapters have indicated, many of the technologies needed for an energy revolution are virtually ready to go. But the pace of change will be heavily influenced by the ability of societies to overcome the policy barriers that remain. Carl Weinberg, the former director of research and development at the Pacific Gas and Electric Company, and Princeton University scientist Robert Williams note: "The rules of the present energy economy were established to favor systems now in place. Not surprisingly, the rules tend to be biased against solar energy." Most of these rules were created decades ago, when the central issue was how to expand fossil fuel use rapidly. The industries that have grown up under the influence of those policies will lobby fiercely to retain them.[4]

The needed policy changes number in the hundreds, but most fall into one of four categories: reducing subsidies for fossil fuels and raising taxes on them to reflect environmental costs, redirecting research and development spending to focus on critical new energy technologies, accelerating investment in the new devices, and channeling international energy assistance to developing countries. Many of the measures will require recasting the role of government, which in the past has been more centrally involved in the energy sector than in almost any other part of the civilian economy. In most areas, greater reliance on the market and less direct government involvement are called for, although in a number of cases governments will need to set the rules, focusing on ways to ensure that environmental costs are considered when economic decisions are made.

*   *   *   *

Energy price reform is a prerequisite to the development of a sustainable energy system. Governments routinely provide heavy subsidies to traditional energy sources, keeping prices artificially low and encouraging waste. In 1991, direct fossil fuel subsidies totalled some $220 billion a year worldwide, according to World Bank estimates, equivalent to 20–25 percent of the value of all fossil fuels sold. China and the formerly centrally planned economies of Europe accounted for three quarters of those subsidies, though the situation is changing rapidly in both areas. At the same time, most developing countries continue long-standing policies that keep energy prices low. Consumers there pay, for example, just 60 percent of what more electricity will cost, with average tariffs dropping from 7.0¢ a kilowatt-hour in 1983 to 4.9¢ by 1988 (in 1993 dollars). World Bank economists estimate that gradually removing such subsidies worldwide would cut carbon emissions in 2010 to 7 percent below the projected level.[5]

In industrial market countries, smaller but still perni-

cious subsidies exist, though instead of being direct
price supports, they generally take the form of tax policies, government loan guarantees, and assured markets.
In the United States, energy industries received federal
subsidies worth more than $36 billion in 1989 (the most
recent data available), with the fossil fuel and nuclear
industries reaping nearly 60 percent and 30 percent, respectively, of the total. Other countries also lavish financial breaks for particular energy sources, from nuclear
power in the United Kingdom to coal in Germany and
hydropower in Quebec, Canada.[6]

Governments have begun to remove some of these
subsidies, occasionally with spectacular results. In the
United Kingdom, for instance, gradual withdrawal of
subsidies to coal—which were channeled through the
government-owned utility company before it was broken up—resulted in a 20-percent decline in coal use between 1990 and 1993, with further erosion expected in
the next few years. Similarly, China has begun to reduce
its sizable coal subsidies.[7]

The next step in energy price reform is to ensure that
fossil fuel prices reflect their full environmental costs.
One of the best ways to incorporate these into day-to-
day economic decisions is to levy energy taxes that are
high enough so that the environmental cost is roughly
embodied in the price consumers pay. A leading propo-
nent of this concept is Ernst von Weizsäcker, president
of the Wuppertal Institute in Germany, who for several
years has argued for a tax system that "tells the ecological truth." Von Weizsäcker proposes gradually increasing taxes on fossil fuels, pushing prices up by 5–7 percent a year until eventually oil costs more than $100 per
barrel. Such a tax would encourage individuals and
companies to choose fuels based on their relative contri-

bution to global warming. Coal would be taxed the highest, oil would be next, and natural gas would follow. Renewable energy sources that do not contribute to the buildup of carbon dioxide would not be taxed at all.[8]

One study of European transport found that the continent's already high gasoline taxes would need to be increased by at least 50 percent to incorporate the full energy and pollution cost of transportation. The impact on consumers could be eased by a partially offsetting reduction in other taxes, while some of the tax receipts could be used to finance new technologies and help poorer developing countries shift to a sustainable energy path. Although such a tax would eventually have to become substantial—perhaps $100 per ton of carbon, for a worldwide total of $590 billion per year—if it were implemented gradually and with offsetting reductions elsewhere, the net economic effect could actually be positive. (For comparison, the World Bank's estimate of the 1989 fossil fuel subsidy amounts to a reverse carbon tax of $40 a ton.) Others have proposed a more complex energy tax that would cover a variety of other pollutants as well as carbon dioxide.[9]

The idea of a tax on carbon dioxide attracted the attention of several governments just before the Earth Summit. At one point, the European Community and Japan both seemed on the brink of adopting one, but they backed away when they were unable to forge an international consensus on the proposal. National leaders realized that if they adopted such taxes unilaterally, they might disadvantage their energy-intensive industries—steel and aluminum, for example. Even the four countries that already have carbon taxes—Denmark, Finland, Norway, and Sweden—have faced internal pressure to drop them. Bucking the trend, Switzerland announced in early 1994 that it would soon introduce a modest carbon tax; at about the same time, Jacques Delors, president of the European Union (which superseded the European Community), called again for an international tax on carbon emissions.[10]

Environmental costs can be internalized in energy decision making in a variety of other ways, some of which were described in earlier chapters. Many environmental laws, for example, force energy users to pay the cost of meeting a pollution standard; in the United States, federal, state, and local governments have taken the additional step of setting a pollution cap for a region and then allowing companies to buy and sell pollution permits. In some cases this has effectively created a market for clean energy and encouraged companies to invest in an array of new energy options. Although these caps sometime create regional inequities, they are an effective spur for new energy technologies. The caps will need to be gradually reduced over time, however, if they are to have continuing positive effects. In addition, costs can be internalized through the electric utility planning process by requiring that they be weighed when different power options are evaluated by utility economists.[11]

The second major area that cries out for reform is the energy R&D programs of governments. Nuclear energy and fossil fuels have traditionally dominated the portfolios of government research efforts. In fact, the 23 member governments of the International Energy Agency spent 85 percent of their $115-billion energy research budgets on nuclear energy and fossil fuels between 1978 and 1991, and less than 6 and 9 percent, respectively, on energy-efficiency and renewable energy technologies. (See Table 14–1.) In most countries, energy R&D budgets continue to reflect priorities of the recent past, not those of the future.[12]

TABLE 14-1. *Government Research and Development Spending in International Energy Agency Member Countries*

Technology	Total, 1978–91		1991	
	Amount	Share	Amount	Share
	(billion 1991 dollars)	(percent)	(billion 1991 dollars)	(percent)
Nuclear Fission	59.8	52	3.6	47
Nuclear Fusion	12.2	11	0.9	12
Gas Turbines[1]	10.9	9	0.6	8
Other Fossil Technologies	14.4	13	1.3	17
Photovoltaics	2.7	2	0.2	3
Other Renewables	7.1	7	0.4	5
End-Use Efficiency	6.6	6	0.6	8
Fuel Cells	1.0	1	< 0.1	1
Total[2]	114.7	100	7.6	100

[1]Mostly spent by defense ministries for aircraft development, but has commercial applications. [2]Columns may not add to totals due to rounding.

SOURCE: See endnote 12.

By the mid-nineties, the tilt toward nuclear power and fossil fuels remained, though spending on efficiency and renewables had begun to rise. In the United States, for example, government-funded R&D in efficiency and renewables climbed 72 and 67 percent, respectively, between 1991 and 1994, with additional increases proposed in President Clinton's 1995 budget. In the past, R&D funds were wasted on large, premature demonstration projects, but most governments now seem to recognize that smaller efforts to advance key technologies and cost-shared commercialization efforts with private companies are more effective. Planners would do well to examine the experience with solar photovoltaics and fuel cells, where government R&D efforts—much of it channeled through military R&D—took key technologies to the threshold of widespread commercial use.[13]

The third priority for energy reformers is to spur the expansion of commercial markets for technologies such as efficient electric motors, wind turbines, fuel cells, and a host of other innovations. Creating large markets will encourage companies to scale up production and reduce the costs of these manufactured devices. Economists who track such price reductions use a "learning curve" to measure the gains. The Boston Consulting Group, in a review of numerous industries, found that each time cumulative production doubled, the average unit price fell 20–30 percent. The classic example of this is the Ford Motor Company's experience with the Model T. Between 1909 and 1923, Ford reduced the price by two thirds as annual production rose from 34,000 to 2.7 million. (See Figure 14–1.) In more recent years, the burgeoning aircraft market was essential to bringing down the cost of gas turbines.[14]

Technologies such as fuel cells, compact fluorescent bulbs, wind turbines, and photovoltaics are still in the early stages of what is likely to be an extended period of cost reduction. Total production of wind turbines, for example—from all companies combined—was less than 2,000 units a year in the early nineties, while the largest manufacturer of fuel cells recently built a facility capable of producing just 200 intermediate-sized cells per year. Some of the most rapid cost gains are likely to come in photovoltaics, which have dropped 33 percent in price for each doubling of cumulative production since 1975. (See Figure 14–2.) The faster these markets grow in the

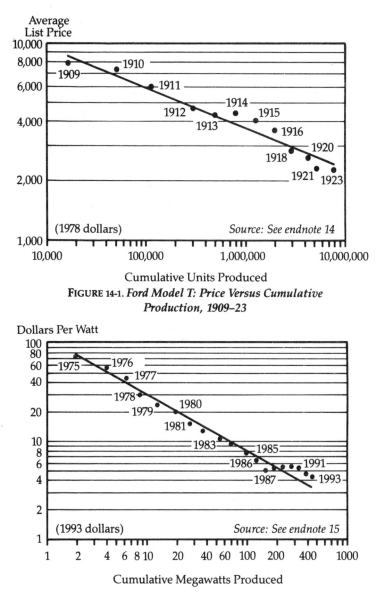

FIGURE 14-1. *Ford Model T: Price Versus Cumulative Production, 1909–23*

FIGURE 14-2. *Photovoltaic Modules: Price Versus Cumulative Production, 1975–93*

future, the more economical the technologies will become.[15]

The key goal for policymakers is to catalyze market-driven, multiyear purchases, so that manufacturers can scale up production. A government strategy to promote continuing development of new technologies can rely either on direct purchasing programs or on partnerships with private industry. These could range from large purchases of solar generators for use on military bases or government buildings to utility demand-side management programs that use corporate capital as leverage to expand the market for energy-efficient lighting, refrigerators, or furnaces.

The government of Japan has aggressively pursued this approach with other technologies in the past, and is now using it to generate a large market for fuel cells and photovoltaics over the next decade. Several European governments are doing the same for photovoltaics. At the same time, a group of U.S. utilities is arranging bulk purchases of solar cells, albeit on a scale that, according to many in the industry, is not large enough to make a significant difference. In addition, some U.S. companies are working with the government to build a series of large solar thermal power plants on a former nuclear test site in the Nevada desert.[16]

To be effective, such programs must be sustained over time, and purchases should be awarded through competitive bids, providing manufacturers with an incentive to reduce costs. Companies that jump in early will still be taking larger risks, but they will stay a step ahead of their competitors on the technological learning curve.

Similar approaches can be used to spur the market for more efficient and less polluting motor vehicles. Already, Austria, Denmark, Germany, Sweden, and the province of Ontario in Canada have vehicle taxes based in part on emissions or fuel economy. Unlike traditional pollution and fuel economy standards, these give consumers incentives to buy better cars rather than just forcing manufacturers to build them. In California, a proposal known as DRIVE+ would tax vehicles whose emissions of carbon dioxide and regulated pollutants rise above certain levels, while a rebate would be available for the purchase of more-efficient or cleaner-running cars. As proposed, DRIVE+ would be revenue-neutral—taxes on dirtier cars would cover the rebates for cleaner ones. An analysis by economists with DRI/McGraw Hill found that such a program would result in a more fuel-efficient car fleet quicker and at less cost than any of the other options studied. Although auto industry lobbyists have so far blocked this measure in California, it enjoys strong support in the legislature. In 1994, a committee of the European Parliament proposed a similar program of incentives and disincentives on new vehicle purchases.[17]

The final key to reform lies in developing countries, where most of the increased demand for energy over the next three decades is expected. Turning the energy market toward new technologies there is critical. Rather than spending billions of dollars on infrastructure that is increasingly obsolete, developing countries can follow a sustainable energy strategy from the start. To do this, however, the large energy assistance programs that earlier pushed them down the fossil fuel path need to be redirected.[18]

The multilateral development banks, especially the World Bank, play the largest role. Although these agencies provide no more than 10 percent of the roughly $60 billion invested in developing countries' energy sectors each year, they often supply the "seal of approval" that private banks are looking for. Nearly four fifths of the $57 billion the World Bank has loaned for energy projects since 1948 has been for power supply; less than 1 percent of the $67 billion lent for energy projects by all the development banks in the eighties went to improving end-use energy efficiency, according to the International Institute for Energy Conservation of Washington, D.C.[19]

Changes in priorities are needed, and at the World Bank they may be afoot. In recent years, for example, the organization has increased its support for natural gas, apparently at the expense of electric power projects. Late in 1991, the Bank formed an alternative energy unit in its Asian section to encourage investments in energy efficiency and renewable energy technologies. And the following year the Bank's directors approved a new policy for the electric power sector, intended to promote energy efficiency. Still, there is little evidence so far that this is having much effect on actual lending patterns; during the first half of 1993, only 2 of the 46 electricity loans being considered were consistent with the new policy.[20]

World Bank staff claim that many of the new energy technologies are untested and too expensive to meet their lending criteria, a problem being addressed through the Global Environment Facility (GEF). As noted in Chapter 8, this fund was set up in 1990 under the auspices of the World Bank, the United Nations Development Programme, and the United Nations Environment Programme to fund projects that are not yet fully justifiable in local economic terms but that have global environmental benefits. In its first three years, the

GEF financed several energy efficiency and renewable energy projects under its mandate to support programs that reduce greenhouse gas emissions. Included were a rooftop photovoltaics project in Zimbabwe, a project for extracting energy from sugarcane residues in Mauritius, and an efficient lighting program in Mexico.[21]

Although some of the GEF energy projects are worthwhile, the effects have been limited so far. The total funding available to GEF—$2 billion for the three years starting in mid-1994, of which 40–50 percent can be energy-related—is far too little to reform energy development. In fact, the multilateral development banks lend 18 times as much each year for traditional energy projects. Moreover, by including energy-efficient lighting, for instance, in its portfolio, GEF gives the false impression that these projects are not economical on their own and should only be pursued for their global environmental benefits—perhaps undermining their economic legitimacy in the eyes of government planners. The institution's potential impact will only be felt if its projects begin to influence the broader energy loan portfolios of the development banks themselves—a point that GEF appears to be addressing in its next round of loans.[22]

GEF's limitations also expose a glaring weakness of the United Nations: the only central U.N. energy agency is the International Atomic Energy Agency—which promotes the export of nuclear power to developing nations in addition to monitoring nuclear proliferation. This one-dimensional organization has been criticized by the World Bank itself for providing poor advice to developing countries. In response, a growing number of critics argue for a broader approach by the international community.[23]

A report completed by a U.N. advisory panel in 1992 recommended the creation of a U.N. solar energy agency that would take the lead in promoting new energy technologies. It would be decentralized, incorporating a series of research stations in key regions around the world. These centers could help demonstrate and commercialize everything from local biogas systems to buildings designed with the local climate in mind. They would be particularly useful to smaller developing countries that do not have the resources of, say, Brazil or India.[24]

*   *   *   *

The policy changes covered in this chapter and earlier in *Power Surge* are not like orders at a Japanese sushi bar—one from each column. Swift and comprehensive action in all four areas is needed. Only then can the world move to an energy system that is not just more benign environmentally but that makes economic sense as well. To be effective, government planners will need to be flexible, abandoning programs that do not work and fine-tuning those that do. And they will need to find ways to harness the power of the market more efficiently, reducing the need for government micromanagement of the energy system.

The road ahead will not be without obstacles and detours. The investments required for achieving a sustainable energy system are sizable, the economic forces to be overcome well organized, and the challenges to human ingenuity enormous. Still, when economic historians look back on the mid-nineties, they may well decide that the world had already embarked on a major energy transition by then—just as, with hindsight, we can say the same about the 1890s. Today, as then, economic, environmental, and social pressures have made the old system unsustainable and obsolete, and the process of change is quietly gathering momentum.

Slowly, people and governments are rising to one of the most fundamental challenges humanity has ever faced: passing on to our children a natural environment that has not been substantially degraded by our short-term needs or desires.

While public fervor about the climate dilemma has been smoldering quietly since the Earth Summit, it could reignite at a moment's notice—sparked by a severe drought, a few hurricanes, or a summer heat wave. Moreover, the world is making rapid progress on many fronts, paving the way for accelerated commercial development of new technologies. It is worth remembering that at similar points in the transition to oil and electricity in the early part of this century, few saw the shift coming. And even those who did failed to appreciate its speed.

Driven by the challenges of creating a less environmentally destructive energy system and lured by the market opportunities that beckon, a new generation of private entrepreneurs, grassroots activists, and policy innovators is already laying the groundwork for an energy transition. Just who will emerge as the Thomas Edison or Henry Ford of the coming energy revolution is unclear, but they are almost certainly out there today—inventing new electric vehicles, assisting villagers install solar lighting systems, and fighting before regulatory commissions to reform the utility industry.

After two decades of mainly uphill battles, the forces of change may finally be gaining ground in the effort to forge a more sustainable energy system. If a persuasive vision of that energy system comes into focus soon, the transition is likely to accelerate—solving energy problems that have plagued humanity for decades, and creating a host of new economic opportunities.

Notes

CHAPTER 1. Power Shift

1. Quote from Alan Kay's office, Apple Computer, Los Angeles, Calif., private communication to Harvey Sachs, Center for Global Change, College Park, Md., July 6, 1992.
2. See, for example, World Energy Council (WEC), *Energy for Tomorrow's World* (New York: St. Martin's Press, 1993), Organisation for Economic Co-operation and Development (OECD), International Energy Agency (IEA), *World Energy Outlook* (Paris: 1993), and U.S. Department of Energy (DOE), Energy Information Administration (EIA), *Annual Energy Outlook 1994 With Projections to 2010* (Washington, D.C.: U.S. Government Printing Office (GPO), 1994).
3. International Atomic Energy Agency, *Annual Reports* (Vienna: 1972–80); Ford Foundation, *Energy: The Next Twenty Years* (Cambridge, Mass.: Ballinger, 1979).
4. WEC, op. cit. note 2.
5. See Chapters 4 and 5 for detailed references.
6. See Chapters 6, 7, and 8 for detailed references.
7. See Chapter 10 for detailed references.

8. U.S. Environmental Protection Agency, *Policy Options for Stabilizing Global Climate* (draft) (Washington, D.C.: 1989); OECD, op. cit. note 2; Intergovernmental Panel on Climate Change (IPCC), *Climate Change: The IPCC Scientific Assessment* (Cambridge, U.K.: Cambridge University Press, 1990); Thomas A. Boden et al., *Trends '93* (Oak Ridge, Tenn.: Oak Ridge National Laboratory, 1994); for detailed figures on carbon emission trends, see Chapter 13.

9. United Nations, *Long-range World Population Projections: Two Centuries of Population Growth, 1950–2150* (New York: 1992); WEC, op. cit. note 2.

10. Worldwatch estimate, based on British Petroleum (BP), *BP Statistical Review of World Energy* (London: 1993) and electronic database, on DOE, EIA, *Annual Energy Review 1992* (Washington, D.C.: GPO, 1993), and on D.O. Hall, King's College London, private communication, February 18, 1994, and printout, March 7, 1994; William Pepper et al., "Emission Scenarios for the IPCC: An Update," prepared for the IPCC Working Group I, May 1992; BP, op. cit. in this note.

11. Verne reference from Peter Hoffmann, *The Forever Fuel: The Story of Hydrogen* (Boulder, Colo.: Westview Press, 1981).

CHAPTER 2. Oil Shock

1. The Sunday Times, *The Yom Kippur War* (London: Andre Deutsch, Ltd., 1975).

2. Daniel Yergin, *The Prize: The Epic Quest for Oil, Money & Power* (New York: Simon & Schuster, 1990).

3. Richard Nixon, "Address to the Nation about Policies to Deal with the Energy Shortages, November 7, 1973," in *Public Papers of the Presidents of the United States—Richard Nixon 1973* (Washington, D.C.: U.S. Government Printing Office (GPO), 1974); details of Nixon's energy policy, which was in the process of being formulated before the oil embargo, can be found in Richard Nixon, "Special Message to the Congress on Energy Policy, April 18, 1973," in ibid.

4. Yergin, op. cit. note 2.

5. Richard Munson, *The Power Makers* (Emmaus, Pa.: Rodale Press, 1985).

6. Worldwatch estimate based on United Nations (UN), *1990 Energy Statistics Yearbook* (New York: 1992), and on Edison Electric Institute (EEI), *Statistical Yearbook of the Electric Utility Industry 1991* (Washington, D.C.: 1992); size of world auto industry is a Worldwatch Institute estimate based on Depart-

ment of Commerce, Bureau of the Census, *Statistical Abstract of the United States 1992* (Washington, D.C.: 1992), on American Automobile Manufacturers Association, *World Motor Vehicle Data 1993* (Detroit, Mich.: 1993), and on Motor Vehicle Manufacturers Association of the United States, Inc., *Facts & Figures '90* (Detroit, Mich.: 1990).

7. Figure 2–1 is based on British Petroleum (BP), *BP Statistical Review of World Energy* (London: 1993) and electronic database (London: 1992), and on Worldwatch estimates, based on ibid., on D.O. Hall, King's College London, private communication, February 18, 1994, and printout, March 7, 1994, and on UN, *World Energy Supplies, 1950–1974* (New York: 1976); biomass estimates for Figure 2–1 before 1970 assume 35 exajoules of use in 1950, compared with 39 in 1970; electricity growth figure based on ibid.

8. Yergin, op. cit. note 2.

9. Peter Schwartz, *The Art of the Long View* (New York: Doubleday Currency, 1991).

10. Yergin, op. cit. note 2; Figure 2–2 is based on BP, op. cit. note 7, and on U.S. Department of Energy (DOE), Energy Information Agency (EIA), *Monthly Energy Review March 1994* (Washington, D.C.: GPO, 1994).

11. Christopher Flavin, *Nuclear Power: The Market Test*, Worldwatch Paper 57 (Washington, D.C.: Worldwatch Institute, December 1983); Christopher Flavin, *Reassessing Nuclear Power: The Fallout From Chernobyl*, Worldwatch Paper 75 (Washington, D.C.: Worldwatch Institute, March 1987); Greenpeace International, WISE-Paris, and Worldwatch Institute, *The World Nuclear Industry Status Report: 1992* (London: 1992).

12. Figure 2–3 is based on Worldwatch database compiled from Mycle Schneider and Assad Kondakji, WISE-Paris, private communication and electronic file, January 14, 1994, from Nuclear Engineering International, *The World Nuclear Industry Handbook* (London: 1994), from "World List of Nuclear Power Plants," *Nuclear News*, March 1994, from International Atomic Energy Agency, *Nuclear Power Reactors in the World* (Vienna: 1993), from Greenpeace International, WISE-Paris, and Worldwatch Institute, op. cit. note 11, and from additional press clipping and private communications; electricity share is from UN, *Energy Statistics Yearbook* (New York: 1991); primary energy share is a Worldwatch estimate, based on BP, op. cit. note 7; Wolf Häfele, "Energy from Nuclear Power," *Scientific American*, September 1990.

13. BP, op. cit. note 7; Christopher Flavin, *World Oil: Coping with*

the *Dangers of Success*, Worldwatch Paper 66 (Washington, D.C.: Worldwatch Institute, July 1985).

14. Nicholas Lenssen, *Empowering Development: The New Energy Equation*, Worldwatch Paper 111 (Washington, D.C.: Worldwatch Institute, November 1992); Nicholas Lenssen, "All the Coal in China," *World Watch*, March/April 1993.

15. Table 2–1 is a Worldwatch estimate, based on BP, op. cit. note 7, on DOE, EIA, *Annual Energy Review 1992* (Washington, D.C.: GPO, 1993), on Hall, op. cit. note 7, and on UN, op. cit. note 12.

16. Figure 2–4 is based on Sam H. Schurr and Bruce C. Netschert, *Energy in the American Economy, 1850–1975* (Baltimore, Md.: Johns Hopkins Press, 1960), on DOE, op. cit. note 15, and on DOE, op. cit. note 10.

17. BP, op. cit. note 7; Aaron Sachs, "Population Increase Drops Slightly," in Lester R. Brown, Hal Kane, and David Malin Roodman, *Vital Signs 1994* (New York: W.W. Norton & Company, 1994); Lester R. Brown, "World Economy Expanding," in ibid.

18. Department of Transportation, *National Transportation Statistics: Annual Report* (Washington, D.C.: 1992); 40–45 percent figure is a Worldwatch estimate, based on UN, *World Energy Supplies 1950–1974* (New York: 1976), on OECD, IEA, *Energy Balances of OECD Countries* (Paris: various years), on DOE, EIA, *Annual Energy Review 1992* (Washington, D.C.: GPO, 1993), on BP, op. cit. note 7, on Robert Summers and Alan Heston, "The Penn World Table (Mark 5): An Expanded Set of International Comparisons, 1950–1988," *Quarterly Journal of Economics*, May 1991, and electronic database, and on International Monetary Fund, *World Economic Outlook*, October 1993.

19. Andrew Hill, "Extracting the Benefits from African Oil," *Financial Times*, May 28, 1992; World Bank, *World Development Report 1993* (New York: Oxford University Press, 1993).

20. U.S. Congress, Office of Technology Assessment, *Energy in Developing Countries* (Washington, D.C.: GPO, 1991); C. Rammanohar Reddy et al., "The Debt-Energy Nexus: A Case Study of India," International Energy Initiative, Bangalore, India, undated.

21. UN, *World Population Prospects: The 1992 Revision* (New York: 1992); Table 2–2 is based on BP, op. cit. note 7, on Hall, op. cit. note 7, and on UN, op. cit. this note.

22. Japan price drop is based on UN, *Statistical Yearbook* (New York: 1976), on IMF, *International Financial Statistics Yearbook* (Washington, D.C.: 1993), and on BP, op. cit. note 7; DOE, op. cit. note 15.

23. BP, op. cit. note 7; Gregory F. Ulmishek and Charles D. Masters, "Oil, Gas Resources Estimated in the Former Soviet Union," *Oil & Gas Journal*, December 13, 1993.

24. Oil figures are for crude oil only, and are from DOE, op. cit. note 10, and from DOE, EIA, *Annual Energy Outlook With Projections to 2010* (Washington, D.C.: 1994).

25. BP, op. cit. note 7; Ray Moseley, "Soviet Oil Industry Suffers Deepest Crisis in its History," *Journal of Commerce*, September 17, 1991; "Worldwide Production Falls as Market Plays its Wild Cards," *Oil & Gas Journal*, December 27, 1993; DOE, op. cit. note 10.

26. Philip Verleger, International Institute of Economics, Washington, D.C., private comunication, November 9, 1993.

27. "Worldwide Production Falls," op. cit. note 25.

28. Ibid.; Figure 2–5 is based on American Petroleum Institute, *Basic Petroleum Data Book* (Washington, D.C.: 1993), and on "World's 1993 Oil Flow Slips; Demand to Move Up in 1994," *Oil & Gas Journal*, March 14, 1994.

29. "Saudi Arabian Sands," *Energy Economist*, November 1993.

30. Ed Morse, "The Coming Oil Revolution," *Foreign Affairs*, Winter 1990–91; BP, op. cit. note 7.

31. Jeff Gerth, "Saudi Stability Hit By Heavy Spending Over Last Decade," *New York Times*, August 21, 1993.

Chapter 3. Eco Shock

1. Michael G. Renner, "Military Victory, Ecological Defeat," *World Watch*, July/August 1991.

2. Hilary F. French, *Clearing the Air: A Global Agenda*, Worldwatch Paper 94 (Washington, D.C.: Worldwatch Institute, January 1990); Christopher Flavin, *Reassessing Nuclear Power: The Fallout From Chernobyl*, Worldwatch Paper 75 (Washington, D.C.: Worldwatch Institute, March 1987); Bryan Hodgson, "Alaska's Big Spill: Can the Wilderness Heal?" *National Geographic*, January 1990.

3. Rio number from U.N. Environment Programme, New York Office, private communication, October 26, 1992; UNCED Secretariat, "154 Signatures on Climate Convention in Rio," press release, June 15, 1992.

4. Daniel Yergin, *The Prize* (New York: Simon & Schuster, 1991); Organisation for Economic Co-operation and Development (OECD), International Energy Agency, *World Energy Outlook* (Paris: 1992).

5. Adam Kahane, "Global Scenarios for the Energy Industry: Challenge and Response," Shell International, London, January

1991; Business Council for a Sustainable Energy Future, "The Sustainability Charter," Washington, D.C., December 11, 1992.

6. Size of enegy system is Worldwatch estimate based on U.S. Department of Energy, Energy Information Agency, *State Energy Price and Expenditure Report 1991* (Washington, D.C.: U.S. Government Printing Office, 1993), and on British Petroleum (BP), *BP Statistical Review of World Energy* (London: 1993).

7. French, op. cit. note 2.

8. James M. Lents and William J. Kelly, "Clearing the Air in Los Angeles," *Scientific American*, October 1993; U.S. Congress, "Clean Air Act Amendment of 1970," Public Law 91–97, December 31, 1970.

9. Lents and Kelly, op. cit. note 8; International Union of Air Pollution Prevention Associations, *Clean Air Around the World: The Law and Practice of Air Pollution Control in 14 Countries in 5 Continents* (Brighton, U.K.: 1988); OECD, *OECD Environmental Reviews: Germany* (Paris: 1993); European Conference of Ministers on Transport, *Transport Policy and the Environment* (Paris: OECD, 1990); Stacy C. Davis and Melissa D. Morris, *Transportation Energy Data Book: Edition 12* (Oak Ridge, Tenn.: Oak Ridge National Laboratory (ORNL), 1992); John McCarthy, National Vehicle and Fuels Emission Laboratory, U.S. Environmental Protection Agency (EPA), Ann Arbor, Mich., private communication and printout, March 30, 1994; Bob Bjorge, General Electric, Schenectady, N.Y., private communication and printouts, August 26, 1993.

10. Lents and Kelly, op. cit. note 8; 86 million figure is from EPA, *National Air Quality and Emissions Trends Report, 1991* (Research Triangle Park, N.C.: 1993); Figure 3–1 is based on Council on Environmental Quality, *Environmental Quality* (Washington, D.C.: Executive Office of the President, various years), and on EPA, Office of Air Quality Planning and Standards, *National Air Pollutant Emissions Trends, 1900–1992* (Research Triangle Park, N.C.: 1993).

11. Exposure to particulates and sulfur dioxide from World Bank, *World Development Report 1992* (New York: Oxford University Press, 1992); auto growth from OECD, *OECD Environmental Data, Compendium 1993* (Paris: 1993).

12. Tod Robberson, "Red-Eyed Mexico City Ends Year Arguing About Smog," *Washington Post*, January 1, 1994; "Vehicular Pollution Makes Breathing Dangerous," *Indian Post* (Bombay), February 11, 1989; Mary Kay Magistad, "Bangkok's Progress Marked by Health Hazards," *Washington Post*, May 7, 1991; lead poisoning from World Bank, op. cit. note 11.

13. Yingzhong Lu, *Fueling One Billion: An Insider's Story of Chinese Energy Policy Development* (Washington, D.C.: The Washington Institute, 1993); Jonathan E. Sinton, ed., *China Energy Databook* (Berkeley, Calif.: Lawrence Berkeley Laboratory, November 1992); William U. Chandler et al., "Energy for the Soviet Union, Eastern Europe and China," *Scientific American*, September 1990; Vaclav Smil, *China's Environmental Crisis: An Inquiry into the Limits of National Development* (New York: M.E. Sharpe Inc., 1993); H. Keith Florig, "The Benefits of Air Pollution Reduction in China," Resources for the Future, Washington, D.C., draft, November 24, 1993.

14. U.S. Congress, "Clean Air Act Amendments of 1990," Public Law 101–349, November 15, 1990; "U.N. Committee Completes Work on Draft for Reductions in Sulfur Dioxide Emissions," *International Environment Reporter*, March 9, 1994; "New Emission Standards for Vehicles," *T&E Bulletin*, December 1993.

15. EPA, Office of Pollution Prevention and Toxics, *1991 Toxic Release Inventory: Public Data Release* (Washington, D.C.: 1993); Chris Neme, *Electric Utilities and Long-Range Transport of Mercury and Other Toxic Air Pollutants* (Washington, D.C.: Center for Clean Air Policy, 1991); Henry S. Cole, Amy L. Hitchcock, and Robert Collins, *Mercury Warning: The Fish You Catch May be Unsafe to Eat* (Washington, D.C.: Clean Water Action, 1992); Anne K. Rhodes, "Technology, Efficient Operation Key Elements in Environmental Strategy," *Oil & Gas Journal*, November 29, 1993.

16. Satinath Sarangi and Carol Sherman, "Piparwar: White Industries' Black Hole," *The Ecologist*, March/April 1993; D.D. Dharmadhikari et al., "Impact of Coal Mining on Heavy Metals in Water," *Asian Environment*, Third Quarter 1992; Peter Chilson, "Coal Miners' Story," *Audubon*, March-April 1994.

17. Bruce Rich, *Mortgaging the Earth: The World Bank, Environmental Impoverishment, and the Crisis of Development* (Boston, Mass.: Beacon Press, 1994); Bruce Rich, Environmental Defense Fund, Washington, D.C., private communication, October 6, 1993.

18. James E. Hansen, NASA Goddard Institute for Space Studies, "The Greenhouse Effect: Impacts on Current Global Temperature and Regional Heat Waves," Testimony before the Committee on Energy and Natural Resources, U.S. Senate, Washington, D.C., June 23, 1988.

19. Sylvia Edgerton, United States Global Change Research Program, Washington, D.C., private communication, January 18, 1994.

20. Intergovernmental Panel on Climate Change (IPCC), *Climate Change: The IPCC Scientific Assessment* (New York: Cambridge University Press, 1990).

21. Svante Arrhenius, "On the Influence of Carbonic Acid in the Air Upon the Temperature of the Ground," *Phil. Mag.*, 1896.

22. H. Friedli et al., "Ice Core Record of the $^{13}C/^{12}C$ Ratio of Atmospheric CO_2 in the Past Two Centuries," *Nature*, November 20, 1986; Figure 3–2 is based on Charles D. Keeling, "Measurements of the Concentration of Atmospheric Carbon Dioxide at Mauna Loa Observatory, Hawaii, 1958–1986," Final Report for the Carbon Dioxide Information and Analysis Center, Martin-Marietta Energy Systems Inc., Oak Ridge, Tenn., April 1987, and on Charles D. Keeling and Timothy Whorf, Scripps Institution of Oceanography, La Jolla, Calif., private communication and printout, February 26, 1993, and February 14, 1994.

23. U.S. National Academy of Sciences, *Changing Climate* (Washington, D.C.: 1983).

24. V. Ramanathan et al., "Trace Gas Trends and Their Potential Role in Climate Change," *Journal of Geophysical Research*, June 20, 1985; Daniel Lashof, Natural Resources Defense Council, Washington, D.C., private communication, January 6, 1994; Figure 3–3 is based on G. Marland and T.A. Boden, "Global, Regional, and National CO_2 Emissions Estimates from Fossil Fuel Burning, Cement Production, and Gas Flaring: 1950–1990" (electronic database) (Oak Ridge, Tenn.: ORNL, 1993), on Thomas Boden, ORNL, Oak Ridge, Tenn., private communication, September 20, 1993, and on BP, op. cit. note 6.

25. IPCC, op. cit. note 20; Richard Kerr, "Pinatubo Global Cooling on Target," *Science*, January 29, 1993.

26. IPCC, op. cit. note 20; IPCC, "Policymakers' Summary of the Potential Impacts of Climate Change," Report of the Working Group II to IPCC (Canberra: Australian Government Publishing Service, 1990); William R. Cline, *Global Warming: The Economic Stakes* (Washington, D.C.: Institute for International Economics, 1992); Figure 3–4 is based on James Hansen and Sergej Lebedeff, "Global Trends of Measured Surface Air Temperature," *Journal of Geophysical Research*, Vol. 92, November 1987, on Helene Wilson and James Hansen, NASA Goddard Institute for Space Studies and Columbia University, New York, private communication, February 18, 1994, and on IPCC, *Climate Change 1992: The IPCC Supplementary Report* (New York: Cambridge University Press, 1992).

27. IPCC, op. cit. note 26; Richard A. Kerr, "Greenhouse Skeptic Out in the Cold," *Science*, December 1, 1989; Richard A. Kerr,

"Greenhouse Science Survives Skeptics," *Science*, May 22, 1992; Patrick Michaels, *Sound and Fury: Science and Politics of Global Warming* (Washington, D.C.: Cato Institute, 1992); Richard Lindzen, "Absence of Scientific Basis," *Research & Exploration*, Spring 1993.

28. Jeremy Leggett, *Climate Change and the Insurance Industry* (Amsterdam: Greenpeace International, 1992); Nancy C. Wilson, "Surge of Hurricanes and Floods Perturb Insurance Industry," *Climate Alert*, July/August 1993; The College of Insurers, "Climate Change and the Insurance Industry: The Next Generation," Proceedings, New York, September 28, 1993.

29. Philippe Sand, "The United Nations Framework Convention on Climate Change: A Preliminary Assessment," Kings College, London, June 2, 1992; IPCC, op. cit. note 20; James C.G. Walker and James F. Kasting, "Effects of Fuel and Forest Conservation on Future Levels of Atmospheric Carbon Dioxide," *Palaeography, Palaeoclimatology, Palaeoecology*, 1992; Interim Secretariat of the United Nations Framework Convention on Climate Change, "Status of Ratification of the United Nations Framework Convention on Climate Change" (electronic bulletin board posting), March 28, 1994.

30. United Nations, "Report of the Intergovernmental Negotiating Committee for a Framework Convention on Climate Change," Fifth Session, Second Part, New York, April 30-May 9, 1992; Patricia Anholt Habr, Treaty Section, Office of Legal Affairs, United Nations, New York, private communication and printout, January 5, 1994.

31. David Malin Roodman, "Pioneering Greenhouse Policy," *World Watch*, July/August 1993.

32. John Courtis, California Air Resources Board, Sacramento, Calif., private communication, April 28, 1994; Public Service Company of New Mexico, "San Juan Generating Station: Fact Sheet," February 20, 1992, cited in Charles Bensinger, "Solar Thermal Repowering: A Technical and Economic Pre-Feasibility Study" (draft revised version 2.0), The Energy Foundation, San Francisco, Calif., 1993; Barbara J. Cummings, *Dam the Rivers, Damn the People* (London: Earthscan Publications Ltd., 1990); Nicholas Lenssen, *Nuclear Waste: The Problem That Won't Go Away*, Worldwatch Paper 106 (Washington, D.C.: Worldwatch Institute, December 1991).

33. Ralph Cavanagh et al., "Utilities and CO_2 Emissions: Who Bears the Risks of Future Regulation?" *The Electricity Journal*, March 1993.

34. "Enter the Btu Tax," *Energy Economist*, May 1993; Ernst U. von

Weizsäcker and Jochen Jesinghaus, *Ecological Tax Reform* (At-
lantic Highlands, N.J.: Zed Books, 1992); Lucy Plaskett, "EC:
Energy Tax," *Energy Economist*, June 1993; David Gardner,
"Japan Shows 'Clear Interest' in Community's Energy Tax
Plans," *Financial Times*, May 20, 1992; "Sweden," *European
Energy Report*, Country Profile, June 1993.

35. Commonwealth of Massachusetts, Department of Public Utili-
    ties, "D.P.U. 91–131," November 10, 1992; Pace University
    Center for Environmental Legal Studies, *Environmental Costs of
    Electricity* (New York: Oceana Publications, 1990); "Survey,"
    Demand-Side Monthly, Gaithersburg, Md., June 1993.

36. "EPA Posts Its Final Rule On Utility Emissions Trading," *Jour-
    nal of Commerce*, March 9, 1993; Barnaby J. Feder, "Sold: $21
    Million of Air Pollution," *New York Times*, March 30, 1993.

37. Michael Miller, "Southern Calif. Clears Way for Trading in
    Pollution," *Journal of Commerce*, January 4, 1994.

38. South Coast Air Quality Management District, *Air Quality
    Management Plan* (Diamond Bar, Calif.: 1989).

39. Davis and Morris, op. cit. note 9.

CHAPTER 4. Doing More with Less

1. Amory B. Lovins, "Energy Strategy: The Road Not Taken?"
   Foreign Affairs, October 1976.

2. Interview with Amory and Hunter Lovins, "The Soft Energy
   Path," *Dartmouth Alumni Magazine*, June 1982.

3. Critics quoted in Amory Lovins and Hugh Nash, *The Energy
   Controversy: Soft Path Questions and Answers* (San Francisco,
   Calif.: Friends of the Earth, 1979).

4. Energy Policy Project of the Ford Foundation, *A Time to Choose:
   America's Energy Future* (Cambridge, Mass.: Ballinger, 1974);
   U.S. Department of Energy (DOE), Energy Information Administration (EIA), *Annual Energy Review 1992* (Washington,
   D.C.: U.S. Government Printing Office (GPO), 1993); U.S.
   energy consumption for 1993 is from DOE, EIA, *Monthly Energy Review March 1994* (Washington, D.C.: GPO, 1994), and
   includes an additional 3 exajoules of energy from biomass; Organisation for Economic Co-operation and Development
   (OECD), International Energy Agency (IEA), *Energy Balances
   of OECD Countries* (Paris: various years).

5. Figure 4–1 is based on DOE, *Annual Energy Review*, op. cit. note
   4, on United Nations (UN), *World Energy Supplies 1950–1974*
   (New York: 1976), and on OECD, op. cit. note 4; 1992 figures
   are Worldwatch estimates based on British Petroleum (BP), *BP*

Statistical Review of World Energy (London: 1993), on OECD, op. cite note 4, and on "German Energy Use Drops But Decline in East Slows," *European Energy Report*, January 7, 1994; GDP data through 1990 from Robert Summers and Alan Heston, "The Penn World Table (Mark 5): An Expanded Set of International Comparisons, 1950–1988," *Quarterly Journal of Economics*, May 1991; 1991 and 1992 GDPs are Worldwatch estimates based on ibid. and on International Monetary Fund (IMF), *World Economic Outlook, October 1993* (Washington, D.C.: 1993); longer-term energy productivity trends are in Amulya K.N. Reddy and José Goldemberg, "Energy for the Developing World," *Scientific American*, September 1990.

6. Angus Maddison, "Growth and Slowdown in Advanced Capitalist Economies: Techniques of Quantitative Assessment," *Journal of Economic Literature*, June 1987.

7. Table 4–1 is based on Robert van der Plas and A.B. de Graaff, "A Comparison of Lamps for Domestic Lighting in Developing Countries," World Bank, Washington, D.C., 1988, and on Mark D. Levine et al., "Electricity End-Use Efficiency: Experience with Technologies, Markets, and Policies Throughout the World," American Council for an Energy-Efficient Economy (ACEEE), Washington, D.C., 1992.

8. Jamie Barrett, Lighting Research Center, Rensselaer Polytechnic Institute, Troy, N.Y., private communication, June 17, 1993; Worldwatch estimate of net present value of the payback from replacing a 60-watt, 1,000-hour incandescent bulb with a 15-watt, 10,000-hour CFL, using a 5-percent annual rate of return on five-year savings, a price of 75¢ for incandescent bulbs and of $20 for CFLs, and a U.S. residential electricity price of 8¢ per kilowatt-hour.

9. Leslie Lammarre, "Shedding Light on the Compact Fluorescent," *EPRI Journal*, March 1993; Evan Mills, Lawrence Berkeley Laboratory, Berkeley, Calif., private communication, February 3, 1993; Nils Borg, Swedish National Board for Industrial and Technical Development (NUTEK), Stockholm, private communication, March 14, 1994.

10. Table 4–2 is based on Howard S. Geller, "Energy-Efficient Appliances: Performance Issues and Policy Options," *IEEE Technology and Society Magazine*, March 1986, on Levine et al., op. cit. note 7, on John Morrill, ACEEE, private communication, May 21, 1993, and on Steven Nadel et al., "Emerging Technologies in the Residential & Commercial Sectors," ACEEE, Washington, D.C., 1993; Figure 4–2 is based on David Goldstein, Natural Resources Defense Council, San Francisco,

      Calif., private communication, June 20, 1993, and on Brooke Stauffer, Association of Home Appliance Manufacturers, private communication, Washington, D.C., December 9, 1993.

11. Nadel et al., op. cit. note 10; "No Pie in the Sky: I.B.M.'s New 'Green' Machine," *New York Times*, May 23, 1993; Otis Port, "Power Surge for a Monster Chip," *Business Week*, September 20, 1993.

12. Deborah Lynn Bleviss, *The New Oil Crisis and Fuel Economy Technologies* (Westport, Conn.: Quorum Books, 1988); Amory B. Lovins, John W. Barnett, and L. Hunter Lovins, "Supercars: The Coming Light-Vehicle Revolution," Rocky Mountain Institute, Snowmass, Colo., 1993.

13. Energy productivity improvement is based on primary energy consumption and is from Lee Schipper and Stephen Meyers, *Energy Efficiency and Human Activity: Past Trends, Future Prospects* (New York: Cambridge University Press, 1992).

14. John E. Barker, American Iron and Steel Institute, letter to Linda G. Stuntz, U.S. Department of Energy, July 31, 1990; R.B. Howarth and L. Schipper, "Manufacturing Energy Use in Eight OECD Countries: Trends through 1988," Lawrence Berkeley Laboratory, Berkeley, Calif., September 1991.

15. G.A. Boyd et al., "Potential for Environmental Impact Mitigation and Energy Conservation in Industry: Case Studies of Steel and Paper Industries" (draft), Argonne National Laboratory, Argonne, Ill., November 8, 1991; Levine et al., op. cit. note 7; Sven Santén, Scan Arc, Hofors, Sweden, private communication, May 27, 1993; Donald F. Barnett and Robert W. Crandall, *Up From the Ashes: The Rise of The Steel Minimill In The United States* (Washington, D.C.: Brookings Institution, 1986).

16. Levine et al., op. cit. note 7; Bill Howe et al., *Drivepower: Technology Atlas* (Boulder, Colo.: E SOURCE, Inc., 1993).

17. Joseph C. Crowley and Bob Sperber, "At KGF, Efficiency Meets 'Environmental Correctness'," *Food Processing*, February 1993.

18. Edison Electric Institute, *Capacity and Generation of Non-Utility Sources of Energy* (Washington, D.C.: various years).

19. Electric Power Research Institute, "Amorphous Metal Transformers Cited as Way to Improve Energy Efficiency," Palo Alto, Calif., news release, April 28, 1993; Allied-Signal Inc., "Improving America's Energy Distribution System," Washington, D.C., undated.

20. U.S. Congress, Office of Technology Assessment, *Green Products by Design* (Washington, D.C.: GPO, 1992); Robert Herman et al., "Dematerialization," in Jesse H. Ausubel and Hedy E.

Sladovich, eds., *Technology and Environment* (Washington, D.C.: National Academy Press, 1989).

21. Schipper and Meyers, op. cit. note 13.

22. Developing countries' energy and economic growth rates exclude China, and are Worldwatch estimates based on Summers and Heston, op. cit. note 5, on IMF, op. cit. note 5, and on BP, op. cit. note 5; Mark Levine et al., *Energy Efficiency, Developing Nations, and Eastern Europe*, A Report to the U.S. Working Group on Global Energy Efficiency (Washington, D.C.: International Institute for Energy Conservation, 1991); Ashok Gadgil et al., "Advanced Lighting and Window Technologies for Reducing Electricity Consumption and Peak Demand: Overseas Manufacturing and Marketing Opportunities," Lawrence Berkeley Laboratory, Berkeley, Calif., March 1991.

23. Mark D. Levine, program leader, Energy and Environment Division, Lawrence Berkeley Laboratory, Berkeley, Calif., private communication, August 19, 1992; Mark D. Levine, Feng Liu, and Jonathan E. Sinton, "China's Energy System: Historical Evolution, Current Issues, and Prospects," *Annual Review of Energy and the Environment*; imports of $53 billion assume that coal and oil contribute equally; Figure 4–3 is based on these sources, on UN, op. cit. note 5, on OECD, IEA, *Energy Balances of Non-OECD Countries 1960–1991* (electronic database) (Paris: 1993), on BP, op cit. note 5, on Summers and Heston, op. cit. note 5, and on IMF, op. cite. note 5.

24. For efficiency potential, see José Goldemberg et al., *Energy for a Sustainable World* (Washington, D.C.: World Resources Institute, 1987), and Levine et al., op. cit. note 22.

25. World Energy Council, *Madrid Statement* (London: 1992); Robert Malpas, "Efficiency, Machiavelli, and Buddah," *Energy: Production, Consumption, and Consequences* (Washington, D.C.: National Academy Press, 1990).

26. Amulya K.N. Reddy, "Barriers to Improvements in Energy Efficiency," Energy and Environment Division, Lawrence Berkeley Laboratory, Berkeley, Calif., October 1991; Amory B. Lovins, "Energy Efficient Buildings: Institutional Barriers and Opportunities," E SOURCE, Inc., Boulder, Colo., December 1992.

27. Ralph Cavanagh et al., "Toward a National Energy Policy," *World Policy Journal*, Spring 1989.

28. Goldstein, op. cit. note 10.

29. Department of Energy Efficiency (DOEE), "Competition Led to Extremely Energy Efficient Fridge-Freeze," NUTEK, Stockholm, undated; Tommy Ankarljung, DOEE, NUTEK, Stockholm, private communication, May 13, 1993.

30. Michael L'Ecuyer et al., "Stalking the Golden Carrot: A Utility Consortium to Accelerate the Introduction of Super-Efficient, CFC-Free Refrigerators," in *ACEEE 1992 Summer Study on Energy Efficiency in Buildings* (Washington, D.C.: ACEEE, 1992); John Feist, Super Efficient Refrigerator Program, Washington, D.C., private communication, April 11, 1994.

31. Jim Harding, "News Flash," in Florentine Krause, "Of Lightbulbs, Lumens and Lamps," *Soft Energy Notes*, May/June 1982; "World List of Nuclear Power Plants," *Nuclear News*, March 1994.

32. Chip Brown, "High Priest of the Low-Flow Shower Heads," *Outside*, November 1991.

Chapter 5. Prince of the Hydrocarbons

1. Wheeler quoted in Robert A. Hefner III, "New Thinking About Natural Gas," in David G. Howell, ed., *The Future of Energy Gases* (Washington, D.C.: U.S. Geological Survey, 1993).

2. Robert A. Hefner III, oral statement to the U.S. Senate, Committee on Energy and Natural Resources, Subcommittee on Energy Regulation, April 26, 1984, as transcribed from audio tape by Janet Rains, GHK Corporation, Oklahoma City, Okla., private communication, April 28, 1994.

3. U.S. Department of Energy (DOE), Energy Information Administration (EIA), *Monthly Energy Review February 1994* (Washington, D.C.: U.S. Government Printing Office (GPO), 1994); British Petroleum (BP), *BP Statistical Review of World Energy* (London: 1993).

4. Howell, op. cit. note 1.

5. Ibid.

6. Jonathan P. Stern, *European Gas Markets* (Aldershot, Hants, U.K.: Royal Institute of International Affairs, 1990); Howell, op. cit. note 1.

7. Gregg Marland, "Carbon Dioxide Emission Rates for Conventional and Synthetic Fuels," *Energy*, Vol. 8, No. 12, 1983; Bob Bjorge, General Electric, Schenectady, N.Y., private communication and printouts, August 26, 1993.

8. Marland, op. cit. note 7; Intergovernmental Panel on Climate Change (IPCC), *Climate Change 1992: The IPCC Supplementary Report* (New York: Cambridge University Press, 1992).

9. United Nations (UN), *World Energy Supplies 1950–1974* (New York: 1976); I.C. Bupp and Frank Schuller, "Natural Gas: Conflicts and Compromise," in Robert Stobaugh and Daniel Yergin, eds., *Energy Future* (New York: Random House, 1979);

Robert L. Bradley, Jr., "Reconsidering the Natural Gas Act," Southern Regulatory Policy Institute Issue Paper No. 5, Roswell, Ga., August 1991.

10. Pietro S. Nivola, "Gridlocked or Gaining Ground?: U.S. Regulatory Reform in the Energy Sector," *The Brookings Review*, Summer 1993.

11. DOE, op. cit. note 3.

12. Nivola, op. cit. note 10; Figure 5–1 is based on DOE, EIA, *Annual Energy Review 1992* (Washington, D.C.: GPO, 1993), and on DOE, op. cit. note 3. Natural gas liquids are counted with natural gas, not crude oil.

13. Caleb Solomon and Robert Johnson, "Natural Gas Industry is Reinventing Itself by Going Global," *Wall Street Journal*, April 19, 1994; developments in various countries from *Energy Economist*, various issues.

14. "Worldwide Construction: Pipelining," *Oil & Gas Journal*, April 11, 1994; Bhushan Bahree, "Siberian Natural Gas Could Solve Energy Needs of Europe," *Wall Street Journal*, April 16, 1993; Daniel A. Dreyfus, "The Pacific Rim and Global Natural Gas," *Energy Policy*, February 1993.

15. Robert Corzine, "The Gas Man Cometh," *Financial Times*, March 22, 1994; "LNG versus an Asian Grid," *Energy Economist*, June 1992.

16. U.S. Bureau of the Census, *Preliminary Data on New Home Construction* (Washington, D.C.: 1994); Organisation for Economic Co-operation and Development (OECD), International Energy Agency (IEA), *Energy Balances of OECD Countries 1990–1991* (Paris: 1993).

17. OECD, IEA, *Energy Policies of IEA Countries: 1991 Review* (Paris: 1992); Richard F. Hirsh, *Technology and Transformation in the American Electric Utility Industry* (New York: Cambridge University Press, 1989).

18. Robert H. Williams and Eric D. Larson, "Expanding Roles for Gas Turbines in Power Generation," in Thomas B. Johansson, Birgit Bodlund, and Robert H. Williams, eds., *Electricity: Efficient End-Use and New Generation Technologies, and Their Planning Implications* (Lund, Sweden: Lund University Press, 1989); Bradley, op. cit. note 9; "The Use of Natural Gas in Power Stations," *Energy in Europe* (Commission of the European Communities, Brussels), December 1990.

19. Eric Jeffs, "First 9F in Service with EdF," *Electricity International*, June/July 1993; Steven Collins, "Special Report: Gas Fired Powerplants," *Power Magazine*, February 1993; Bjorge, op. cit. note 7; "For the Record," *Energy Economist*, April 1993;

"For the Record," *Energy Economist*, September 1993; Neil
Buckley, "Hurdles in the Path of the Dash for Gas," *Financial
Times*, December 10, 1992; the Teeside station has eight gas
turbines and two steam turbines, according to Kristin Rankin,
Enron Corp, Houston, Tex., private communication, October
20, 1993.

20. Williams and Larson, op. cit. note 18; Bjorge, op. cit. note 7;
William H. Day and Ashok D. Rao, "FT4000 HAT with Natural Gas Fuel," *Turbomachinery International*, January/February
1993; Jack Janes, California Energy Commission, Sacramento,
Calif., private communication, May 31, 1994; Ecodyne, Inc.,
"The ACRE Engine: A Revolutionary Technology for the Generation of Electricity," Bellevue, Wash., May 31, 1994; Steven
Collins, "Small Gas Turbines Post Gains in Performance,"
Power Magazine, October 1992; Preben Maegaard, Folkecenter
for Renewable Energy, Hurup Thy, Denmark, private communication, April 21, 1994.

21. Table 5-1 is based on Bjorge, op. cit. note 7, on M.W. Horner,
"GE Aeroderivative Gas Turbines—Design and Operating Features," GE Aircraft Engines, GE Power Generation, Evendale,
Ohio, 1993, on John C. Trocciola, International Fuel Cells,
South Windsor, Conn., private communication, January 3,
1994, and on David S. Bazel, Asea Brown Boveri, North Bruns-
wick, N.J., private communication and printout, November 2,
1993; D.L. Chase, J.M. Kovacik, and H.G. Stoll, "The Economics of Repowering Steam Power Plants," General Electric
Company, Schenectady, N.Y. 1992; Douglas M. Todd and
Robert M. Jones, General Electric, "Advanced Combined Cycles Provide Economic Balance for Improved Environmental
Performance," presented at International Power Generation
Conference, San Diego, Calif., October 6–10, 1991; William
Keeling, "Indonesia's Power Scramble," *Financial Times*, August 10, 1993; GE Power Generation, "MS9001E Gas Turbines: Heavy-Duty 50 Hz Power Plant," Schenectady, N.Y.,
1991; "Europe's Most Modern Combined-Cycle Plant," *Electricity International*, June/July 1993.

22. Brooke Stoddard, "Fuel Cell Update," *American Gas*, June
1993.

23. Fuel Cell Commercialization Group, "What Is a Fuel Cell?"
Washington, D.C., 1992; Philip H. Abelson, "Applications of
Fuel Cells" (editorial), *Science*, June 22, 1990; F.S. Kemp and
J.C. Trocciola, International Fuel Cells, South Windsor, Conn.,
private communication, November 11, 1993.

24. John Douglas, "Utility Fuel Cells in Japan," *EPRI Journal*, Sep-

tember 1991; "SoCalGas Inks First Commercial Fuel Cell Deal," *Oil & Gas Journal*, April 15, 1991.

25. Daniel Sperling, ed., *Alternative Transportation Fuels: An Environmental and Energy Solution* (New York: Quorum Books, 1989); Edward Keller, "International Experience with Clean Fuels," *Fuel Reformulation*, November/December 1993.

26. Michael Walsh, automotive emissions consultant, Washington, D.C., private communication, September 13, 1991; Bryan D. Willson, Colorado State University, Fort Collins, Colo., private communication, September 17, 1991.

27. Alan Caminiti, United Parcel Service, Greenwich, Conn., private communication, September 27, 1991; Alan Pell Crawford, "Choo-Choosing Natural Gas," *American Gas*, September 1993.

28. Neil Geary, spokesperson, Amoco Company, Chicago, Ill., private communication, October 21, 1991; Jessie Ochoa, public affairs officer, Shell Oil Company, Houston, Tex., private communication, October 21, 1991; American Gas Association (AGA), Planning and Analysis Group, "An Analysis of the Economic and Environmental Effects of Natural Gas as an Alternative Fuel," Washington, D.C., December 15, 1991; Nancy Umbach Etkin, Natural Gas Vehicle Coalition, Arlington, Va., private communication, April 28, 1994.

29. Environmental Protection Agency (EPA), "Analysis of the Economic and Environmental Effects of Compressed Natural Gas as a Vehicle Fuel," Washington, D.C., April 1990; Jeffrey A. Alson et al., "Motor Vehicle Emission Characteristics and Air Quality Impacts of Methanol and Compressed Natural Gas," in Sperling, op. cit. note 25; California Air Resources Board, "ARB Certifies First Ultra Low Emission Vehicle," press release, January 7, 1994; AGA, op. cit. note 28.

30. Walsh, op. cit. note 26; Willson, op. cit. note 26.

31. Thomas W. Lippman, "More Use of Natural Gas as Motor Fuel Explored," *Washington Post*, February 14, 1990; Allen R. Wastler, "Calif. Natural Gas Supplier Expands Fuel Station Network," *Journal of Commerce*, March 12, 1991; DOE, EIA, *Monthly Energy Review September 1991* (Washington, D.C.: GPO, 1991); AGA Planning and Analysis Group, "Projected Natural Gas Demand from Vehicles under the Mobile Source Provisions of the Clean Air Act Amendments," Washington, D.C., January 30, 1991.

32. U.S. National Research Council, *Undiscovered Oil and Gas Resources* (Washington, D.C.: National Academy Press, 1991).

33. Robert A. Hefner III, "Natural Gas Resource Base and Produc-

tion Capability Policy Issues," presented at the Aspen Institute Energy Policy Forum, Aspen, Colo., July 13, 1991; AGA Planning and Analysis Group, "Coalbed Methane Resource, Reservoir and Production Characteristics," Issue Brief, Washington, D.C., November 16, 1990.

34. W.L. Fisher, "Factors in Realizing Future Supply Potential of Domestic Oil and Natural Gas," presented at the Aspen Institute Energy Policy Forum, Aspen, Colo., July 13, 1991; Paul D. Holtberg, Gas Research Institute, "Is There Enough Gas?" presented at Americans for Energy Independence seminar on the natural gas outlook, Washington, D.C., October 9, 1991; National Petroleum Council, *The Potential for Natural Gas in the United States: Executive Summary* (Washington, D.C.: 1992); The Potential Gas Committee, *A Comparison of Ultimately Recoverable Quantities of Natural Gas in the United States* (Golden, Colo.: Colorado School of Mines, 1993). It is important to distinguish between proven reserve figures—oil and gas that have been clearly identified through exploratory drilling—and the far greater estimates of ultimately recoverable resources. Figure 5–2 is based on National Petroleum Council, op. cit. in this note, on DOE, op. cit. note 12, and on DOE, op. cit. note 3. Projections were constructed as described in note 41, except that for crude oil, the total ultimately recoverable resource was not estimated and imposed as a constraint during the curve fitting.

35. BP, *BP Statistical Review of World Energy* (London: 1991); Edwin Moore and Enrique Crousillat, "Prospects for Gas-Fueled Combined Cycle Power Generation in the Developing Countries," Energy Series Paper No. 35, World Bank, Washington, D.C., 1991; Shell International Petroleum Company, *Natural Gas* (London: 1988).

36. BP, op. cit. note 3; Gregory F. Ulmishek and Charles D. Masters, "Oil, Gas Resources Estimated in the Former Soviet Union," *Oil & Gas Journal*, December 13, 1993.

37. Table 5–2 is based on Moore and Crousillat, op. cit. note 35, and on UN, *1989 Energy Statistics Yearbook* (New York: 1991); C.D. Masters et al., "Resource Constraints in Petroleum Production Potential," *Science*, July 12, 1991; World Bank, *Annual Report 1991* (Washington, D.C.: 1991); W.W. Crook, "The Potential Impact of Mexico on the North American Gas Market," unpublished report to the National Petroleum Council, Washington, D.C., May 13, 1992.

38. Robert A. Hefner III, "Onshore Natural Gas in China," presented at the World Bank Energy Roundtable Discussion on Gas Development in Less Developed Countries, Paris, March 25–26, 1985.

39. Charles D. Masters, David H. Root, and Emil D. Attanasi, "World Petroleum Assessment and Analysis" (draft), U.S. Geological Survey, Reston, Va., January 1994. The resource figures used here include natural gas liquids such as propane and ethane, fuels that are sometimes neglected in reporting of world gas supplies.

40. Hefner, op. cit. note 1; Thomas Gold, "The Origin of Methane in the Crust of the Earth," in Howell, op. cit. note 1.

41. M. King Hubbert, "Techniques of Prediction as Applied to the Production of Oil and Gas," in *Oil and Gas Supply Modeling*, Department of Commerce Symposium, Washington, D.C., June 18–20, 1980. Gas resource production and reserves in Figure 5–3 include natural gas liquids, and are based on Joel Darmstadter, Perry D. Teitelbaum, and Jaroslav G. Polach, *Energy in the World Economy* (Baltimore, Md.: Johns Hopkins University Press, 1971), on DOE, op. cit. note 12, on American Petroleum Institute, *Basic Petroleum Data Book* (Washington, D.C.: 1993), on UN, op. cit. note 9, on Ray Thomas, Department of Natural Resources, Ottawa, Ont., Canada, private communication, February 23, 1994, on Matthew Sagers, PlanEcon, Inc., Washington, D.C., private communications, February 24, 1994, and on DOE, op. cit. note 3. Unexploited natural gas liquids resource is estimated to be equal to a percentage of the natural gas resource, chosen based on historical relative production rates in the United States. Projections were achieved by fitting the derivative of a "logistic" or "S" curve to historical data using the least-squares metric, while fixing the total area under the curve to given ultimate resource estimates.

42. Worldwatch estimates based on gas resource estimates cited in this chapter and on the Hubbert curve analysis explained above. These figures are consistent with a growth rate in world gas production of 3.5 percent per year during the next two decades, which is about the recent average.

43. Figure of 10 percent is a Worldwatch estimate based on R.M. Rotty and G. Marland, "Production of CO_2 From Fossil-Fuel Burning" (electronic database) (Oak Ridge, Tenn.: Oak Ridge National Laboratory (ORNL), 1993), and on G. Marland and T.A. Boden, "Global, Regional, and National CO_2 Emission Estimates from Fossil Fuel Burning, Cement Production, and Gas Flaring: 1950–1990" (electronic database) (Oak Ridge, Tenn.: ORNL, 1993).

44. IPCC, op. cit. note 8.

45. EPA, *Methane Emissions and Opportunities for Control: Workshop Results of Intergovernmental Panel on Climate Change* (Washington, D.C.: 1990); Daniel Lashof, "Draft Statement on Meth-

ane," Natural Resources Defense Council, Washington, D.C., September 3, 1991.

46. Richard A. Kerr, "Methane Increase Put on Pause," *Science*, February 11, 1994.

47. Nebojsa Nakicenovic, "Energy Gases—The Methane Age and Beyond," in Howell, op. cit. note 1.

CHAPTER 6. Winds of Change

1. Preben Maegaard and René Karottki, Folkecenter for Renewable Energy, "Sustainable Energy Development—the Need for a Decentralized Energy System," presented to the UNESCO Round Table, Strategic Energy Issues in China, Beijing, February 16–18, 1993.

2. René Karottki, Forum for Energy and Development, Copenhagen, private communication, March 9, 1994.

3. Maegaard and Karottki, op. cit. note 1.

4. Birger Madsen, BTM Consult ApS, "Technological Development of Commercial Wind Turbines and Business Opportunities," presented to Seminar on Wind Energy in Southern Europe, Cádiz, Spain, November 11–13, 1993; Birger Madsen, BTM Consult ApS, Ringkoeping, Denmark, private communication to René Karottki, Forum for Energy and Development, Copenhagen, March 21, 1994; 10-percent goal is based on Danish Ministry of Energy, *Energy 2000: A Plan of Action for Sustainable Development* (Copenhagen: 1990).

5. California study is from Mike Batham, California Energy Commission, Sacramento, Calif., private communication, April 7, 1994.

6. Paul Gipe, Gipe and Associates, Tehachapi, Calif., private communication and printout, April 6, 1994.

7. Ibid.

8. Christopher Flavin, *Wind Power: A Turning Point*, Worldwatch Paper 45 (Washington, D.C.: Worldwatch Institute, July 1981).

9. Madsen, private communication, op. cit. note 4.

10. R. Lynette, "Assessment of Wind Power Station Performance and Reliability," EPRI Report GS-6256, Electric Power Research Institute, Palo Alto, Calif., 1989; R. Davidson, "Performance Up and Costs Down," *Windpower Monthly*, November 1991.

11. Susan Hock, Robert Thresher, and Tom Williams, "The Future of Utility-Scale Wind Power," in Karl W. Boer, *Advances in Solar Energy: An Annual Review of Research and Development*, Vol. 7 (Boulder, Colo.: American Solar Energy Society, 1992).

12. J.C. Chapman, *European Wind Technology* (Palo Alto, Calif.: Electric Power Research Institute, 1993); Hock, Thresher, and Williams, op. cit. note 11; David Milborrow, "Variable Speed Comes of Age," *Windpower Monthly*, December 1993.

13. California capacity from Ros Davidson, "Five Hundred Megawatt Agreement Signed," *Windpower Monthly*, April 1994; Sam McMurtrie, American Wind Energy Association (AWEA), Washington, D.C., private communication and printout, February 25, 1994; AWEA, "Wind Energy Industry Sets Strategic Plan for Industry to Install 10,000 MW of Wind by the Year 2000," Washington, D.C., August 17, 1993.

14. Chapman, op. cit. note 12.

15. "The Quixotic Technology," *The Economist*, November 14, 1992; Lyn Harrison, "Europe Gets Clean Away," *Windpower Monthly*, September 1992.

16. Jeff Pelline, "Bay Firm to Supply Windmills to Speed Chernobyl Closure," *San Francisco Chronicle*, February 19, 1993; S. Gopikrishna Warrier, "Wind Power Projects Push Land Cost Sky-High," *Down to Earth*, April, 15, 1993; "Mexico and Chile Follow Suit," *Windpower Monthly*, October 1993; "Argentina Announces Huge Wind Development Project," *Windpower Monthly*, October 1993; Robin Bromby, "U.S. Windpower, Wing Merrill to Build N.Z.'s First Wind Farm," *The Solar Letter*, August 6, 1993.

17. Paul Gipe, Gipe and Associates, Tehachapi, Calif., private communication and printout, January 28, 1994; Jon G. McGowan, "Tilting Toward Windmills," *Technology Review*, July 1993; U.S. Department of Energy, "Technology Evolution for Wind Energy Technology," unpublished draft, June 7, 1993; Figure 6–1 based on Gipe, op. cit. note 6; Leslie Lamarre, "A Growth Market in Wind Power," *EPRI Journal*, December 1992.

18. Henry Kelly and Carl J. Weinberg, "Utility Strategies for Using Renewables," in Thomas B. Johansson, Birgit Bodlund, and Robert H. Williams, eds., *Renewable Energy: Sources for Fuels and Electricity* (Washington, D.C.: Island Press, 1993).

19. Michael J. Grubb and Niels I. Meyer, "Wind Energy: Resources, Systems and Regional Strategies," in Johansson, Bodlund, and Williams, op. cit. note 18; Kelly and Weinberg, op. cit. note 18; Carl Weinberg, Weinberg and Associates, Walnut Creek, Calif., private communication, April 13, 1994.

20. M.N. Schwartz, D.L. Elliott, and G.L. Gower, Pacific Northwest Laboratory, "Seasonal Variability of Wind Electric Potential in the United States," presented at AWEA's Windpower '93 Conference, San Francisco, Calif., July 12–16, 1993; Don

Smith and Mary Ilyin, Pacific Gas and Electric, "Wind and Solar Energy: Costs and Value," *Proceedings of ASME 10th Wind Energy Symposium*, Houston, January 1991.

21. A.J. Cavallo and D.M. Riley, "The Winds of the U.S. Great Plains: A Reevaluation," unpublished, Center for Energy and Environmental Studies, Princeton University, Princeton, N.J., 1993. According to Cavallo and Riley, a reevaluation of wind data for the Great Plains suggests that winds are steadier than previously thought, raising the potential capacity factor. In addition, by adding more wind turbines to a given wind farm and limiting the peak output, the capacity factor can be further increased.

22. Grubb and Meyer, op. cit. note 19; D.L. Elliott, L.L. Windell, and G.L. Gower, *An Assessment of the Available Windy Land Area and Wind Energy Potential in the Contiguous United States* (Richland, Wash.: Pacific Northwest Laboratory, 1991); these figures are for Class 3 wind areas and above, which includes area with a wind power density at a height of 50 meters of 300–400 watts per square meter and an average annual wind speed of at least 6.4 meters per second (14 miles per hour); Table 6–1 is based on Gipe, op. cit. note 6, on Elliott, Windell, and Gower, op. cit. in this note, on Grubb and Meyer, op. cit. note 19, on Edison Electric Institute, "EEI Pocketbook of Electric Utility Industry Statistics," Washington, D.C., 1992, and on Organisation for Economic Co-operation and Development, International Energy Agency, *Energy Policies of IEA Countries: 1992 Review* (Paris: 1993).

23. Grubb and Meyer, op. cit. note 19.

24. Elliott, Windell, and Gower, op. cit. note 22; William Babbitt, Associated Appraisers, Cheyenne, Wyo., private communication, October 11, 1990; Paul Gipe, "Wind Energy Comes of Age," Paul Gipe and Associates, Tehachapi, Calif., May 13, 1990.

25. "UK Wind Power Expansion: The Good, the Bad, and the Ugly?" *Renewable Energy Report*, Supplement to the *European Energy Report*, December 10, 1993; "A New Way to Rape the Countryside," *The Economist*, January 22, 1994; "Wind Energy and the Landscape," *Landscape Architecture*, November 1993.

26. California Energy Commission, "Avian Mortality at Large Wind Energy Facilities in California: Identification of a Problem," Sacramento, Calif., August 1989; Anthony Luke, with Alicia Watts Hosmer and Lyn Harrison, "Bird Deaths Prompt Rethink on Wind Farming in Spain," *Windpower Monthly*, February 1994.

27. Ros Davidson, "Environmentalists Say Stop," *Windpower Monthly*, October 1993; Luke, op. cit. note 26.
28. Resource geography is from Elliott, Windell, and Gower, op. cit. note 22.
29. Richard Stone, "Polarized Debate: EMFs and Cancer," *Science*, December 11, 1992; J. Douglas, "The Delivery System of the Future," *EPRI Journal*, October/November 1992.
30. Alfred J. Cavallo, "Wind Energy: Current Status and Future Prospects," in *Science and Global Security*, Vol. 4, 1993, pp. 1–45.

CHAPTER 7. Heat from the Sun

1. Josef Nowarski, Division of Research and Development, Ministry of Energy and Infrastructure, Jerusalem, private communication, January 19, 1994; Eddie Bet Hazavdi, Energy Conservation Division, Ministry of Energy and Infrastructure, Jerusalem, private communication and printout, January 26, 1994.
2. Ken Butti and John Perlin, *A Golden Thread: 2500 Years of Solar Architecture and Technology* (Palo Alto, Calif.: Cheshire Books, 1980).
3. Ibid.
4. Data for 1994 from Nowarski, op. cit. note 1, and from Bet Hazavdi, op. cit. note 1; Harry Tabor, "Forty Years of Solar Energy Development and Exploitation in Israel," *SunWorld*, March 1993.
5. Solar resource figure from Denis Hayes, *Rays of Hope: The Transition to a Post-Petroleum World* (New York: W.W. Norton & Company, 1977); fossil fuel estimate based on Charles D. Masters, David H. Root, and Emil D. Attanasi, "World Resources of Crude Oil and Natural Gas," *Proceedings of the Thirteenth World Petroleum Congress* (Chichester, U.K: John Wiley & Sons, 1991), on Hans H. Landsberg et al., *Energy: The Next Twenty Years* (Cambridge, Mass.: Ballinger, 1979), and on Jeanne Anderer, Alan McDonald, and Nebojsa Nakicenovic, *Energy in a Finite World* (Cambridge, Mass.: Ballinger, 1981).
6. Vincent E. McKelvey, "Solar Energy in Earth Processes," *Technology Review*, April 1975; 40 percent figure is Worldwatch estimate, based on U.S. Department of Energy (DOE), Energy Information Administration (EIA), *Annual Energy Review 1992* (Washington, D.C.: U.S. Government Printing Office (GPO), 1993), on DOE, EIA, *Monthly Energy Review September 1993* (Washington, D.C.: GPO, (1993), on DOE, EIA, *Annual Energy Outlook 1994* (Washington, D.C.: GPO, 1994), on

Mohammad Adra, EIA, DOE, Washington, D.C., private communication, March 7, 1994, and on "Market Potential for Solar Thermal Energy Supply Systems in the United States Industrial and Commercial Sectors: 1990–2030," Mueller Energy Technology Group (Arlington, Va.: 1991); estimate counts only 20 percent of fuel used in road vehicles and 33 percent in other vehicles as "delivered."

7. Ken Bossong, Sun Day 1994, Takoma Park, Md., private communication, November 19, 1993; Richard Boeth, with William J. Cook, "All Hail the Sun!" *Newsweek*, May 15, 1978.

8. Low-temperature heat is Worldwatch estimate for the United States, based on DOE, *Annual Energy Outlook 1994*, op. cit. note 6, and on Adra, op. cit. note 6.

9. Butti and Perlin, op. cit. note 2; Solar Energy Industries Association, "Solar Thermal Water Heating," Washington, D.C., 1993; Blaine Collison, Passive Solar Industries Association, Washington, D.C., private communication, November 16, 1993.

10. Systems in Japan from Solar System Development Association, "The Status of Solar Energy Systems in Japan," Tokyo, 1993; E.H. Lysen, Netherlands Agency for Energy and the Environment, "Solar Energy in the Netherlands," presented to the IEA-SHCP National Programs Workshops, Sydney, Australia, May 4, 1993.

11. Botswana from Chris Neme, Memorandum to Mark Levine, Lawrence Berkeley Laboratory, Berkeley, Calif., March 28, 1992; Mario Calderón and Paolo Lugari, Centro Las Gaviotas, Bogota, Colombia, private communication, April 13, 1992; Kenya from Christopher Hurst, "Establishing New Markets for Mature Energy Equipment in Developing Countries; Experience with Windmills, Hydro-Powered Mills and Solar Water Heaters," *World Development*, Vol. 18, No. 4, 1990; Jennie Swann, "Solar Bizz in Malawi," *Our Planet*, Vol. 4, No. 6, 1992; Fawwaz Elkarmi and Isam Mustafa, "Increasing the Utilization of Solar Energy Technologies (SET) in Jordan," *Energy Policy*, September 1993.

12. Flat-plate collector costs from Ronald Larson, Frank Vignola, and Ron West, *Economics of Solar Energy Technologies* (Boulder, Colo.: American Solar Energy Society, 1992); natural gas, electricity, and propane costs from DOE, EIA, *Annual Energy Review 1993* (Washington, D.C.: GPO, 1993).

13. James Bergquam, "A Hybrid Solar Absorption Air Conditioning System," *Solar Today*, July/August 1993.

14. Donald Osborn, Sacramento Municipal Utility District, Sac-

ramento, Calif., private communication and printout, April 6, 1994.

15. Nicholas Lenssen, "Cooked by the Sun," *World Watch*, March/April 1989; David A. Ciochetti and Robert H. Metcalf, "Pasteurization of Naturally Contaminated Water with Solar Energy," *Applied and Environmental Microbiology*, February 1984; Robert Metcalf, California State University, Sacramento, private communication and printout, March 16, 1994; number of diarrhoeal deaths from UNICEF, *State of the World's Children 1994* (New York: Oxford University Press, 1994).

16. Lenssen, op. cit. note 15; Darwin O'Ryan Curtis, "An Appreciation of Solar Cooking," SY Dimensions Inc., Healdsburg, Calif., May 1991.

17. Koshy Cherail, "Cloudy Days for Solar Cooker," *Down to Earth*, June 30, 1992; China from Beverly Blum, Solar Box Cookers International, Sacramento, Calif., private communication and printout, July 15, 1993; Pakistan from Peter W. Forbes, "Introduction of Solar Cookers to Rural Refugees," in Solar Energy Society of India, *Renewable Energy for Rural Development* (New Delhi: Tata McGraw-Hill, 1989), with an update from ibid.

18. Mark J. Skowronski et al., Southern California Edison, "The Solar Two Project," presented to Pacific Coast Electrical Association Engineering and Operating Conference, Irvine, Calif., March 18, 1993.

19. Butti and Perlin, op. cit. note 2.

20. François Pharabod and Cédric Philibert, *LUZ Solar Power Plants: Success in California and Worldwide Prospects* (Paris: Agency for the Control of Energy and the Department of the Environment, 1991).

21. Pat De Laquil III et al., "Solar-Thermal Electric Technology," in Thomas B. Johansson, Birgit Bodlund, and Robert H. Williams, eds., *Renewable Energy: Sources for Fuel and Electricity* (Washington, D.C.: Island Press, 1993); David Kearney, KJC Operating Company, Kramer Junction, Calif., private communication, July 30, 1993; Peggy Sheldon, Luz International Limited, Los Angeles, Calif., private communication and printout, August 28, 1990; Don Logan, Luz International Limited, Los Angeles, Calif., private communication, September 26, 1990; Bureau of the Census, U.S. Department of Commerce, *Statistical Abstract of the United States 1990* (Washington, D.C.: GPO, 1990); Figure 7–1 is based on Kearney and on Sheldon, both op. cit. in this note.

22. Number of plants and capacity from David Mills and Bill Kee-

pin, "Baseload Solar Power: Near-term Prospects for Load Following Solar Thermal Electricity," *Energy Policy*, August 1993; export ranking from Michael Lotker, "Barriers to Commercialization of Large-Scale Solar Electricity: Lessons for the LUZ Experience," Sandia National Laboratories, Albuquerque, N.M., November 1991.

23. Douglas N. Koplow, *Federal Energy Subsidies: Energy, Environmental, and Fiscal Impacts* (Washington, D.C.: Alliance to Save Energy, 1993); Newton D. Becker, chairman, Luz International Limited, Testimony before the U.S. House of Representatives, Committee on Ways and Means, February 10, 1992; Lotker, op. cit. note 22; Newton Becker, "The Demise of Luz: A Case Study," *Solar Today*, January/February 1992.

24. Gabi Kennen, Solel, Jerusalem, private communication and printout, June 24, 1993; Becker, op. cit. note 23; David Mills, University of Sydney, private communication and printout, April 5, 1994.

25. Mills and Keepin, op. cit note 22.

26. Ibid.; Mills, op. cit. note 24.

27. Idaho National Engineering Laboratory et al., *The Potential of Renewable Energy: An Interlaboratory White Paper*, prepared for the Office of Policy, Planning and Analysis, DOE, in support of the National Energy Strategy (Golden, Colo.: Solar Energy Research Institute, 1990); Charles Bensinger, "Solar Thermal Repowering: A Technical and Economic Pre-Feasibility Study," The Energy Foundation, San Francisco, Calif., draft, revised version 2.0, 1993.

28. DOE, Office of Solar Energy Conversion, "Solar Thermal Electric Technology Rationale," Washington, D.C., August 1990; Stephen Kaneff, "Mass Utilization of Solar Thermal Energy," Energy Research Centre, Australian National University, Canberra, September 1992.

29. Isoroku Kubo, Cummins Power Generation, Inc., "Cummins Power Generation Dish-Stirling Program," presented at Soltech '93, Washington, D.C., April 27, 1993; Bensinger, op. cit. note 27.

30. De Laquil et al., op. cit. note 21.

31. James E. Cavanagh, John H. Clarke, and Roger Price, "Ocean Energy Systems," in Johansson, Bodlund, and Williams, op. cit. note 21; William J. Broad, "Ocean Pioneer Mines Energy that is Cool, Clean and Free," *New York Times*, July 13, 1993.

32. Bensinger, op. cit. note 27; New Mexico State Legislature, Senate Memorial 45, First Session 1993, Santa Fe; De Laquil et al., op. cit. note 21.

33. Figure of one quarter is Worldwatch estimate based on John P. Holdren, "The Transition to Costlier Energy," prologue, in Lee Schipper and Steven Meyers, *Energy Efficiency and Human Activity: Past Trends and Future Prospects* (New York: Cambridge University Press, 1992) for solar resource, on James J. MacKenzie, Roger C. Dower, and Donald D.T. Chen, *The Going Rate: What It Really Costs to Drive* (Washington, D.C.: World Resources Institute, 1992) for U.S. paved area (which is doubled for a conservative estimate of global paved area), and on British Petroleum, *BP Statistical Review of World Energy* (London: 1993), and D.O. Hall, King's College London, private communication and printout, March 7, 1994, for global primary energy consumption; for pricing, see DOE, op. cit. note 28, Mills and Keepin, op. cit. note 22, and David Mills, University of Sydney, private communication and printout, April 14, 1994.

CHAPTER 8. Plugging into the Sun

1. Hidalgo experience is from Eridania Morales, ADESOL, Bella Vista, Dominican Republic, private communication, February 7, 1992, and from Richard Hansen, Enersol Associates Inc., Somerville, Mass., private communication, February 7, 1992.
2. Figure of more than 2 billion is from Derek Lovejoy, "Electrification of Rural Areas by Solar PV," *Natural Resources Forum*, May 1992.
3. For the history of PV development, see Christopher Flavin, *Electricity from Sunlight: The Future of Photovoltaics*, Worldwatch Paper 52 (Washington, D.C.: Worldwatch Institute, December 1982), and Ken Butti and John Perlin, *A Golden Thread: 2500 Years of Solar Architecture and Technology* (Palo Alto, Calif.: Cheshire Books, 1980).
4. Flavin, op. cit. note 3.
5. Ibid.
6. Ken Zweibel, "Thin-Film Photovoltaic Cells," *American Scientist*, July/August 1993.
7. Figure 8–1 is based on Paul Maycock, Photovoltaic Energy Systems, Inc., Casanova, Va., private communications and printouts, May 8, 1992, and December 20, 1993.
8. Figure 8–2 is based on Paul Maycock, Photovoltaic Energy Systems, Inc., Casanova, Va., private communications, December 20, 1993, and March 23, 1994; for history of cost reductions, see Ken Zweibel, *Harnessing Solar Power: The Photovoltaics Challenge* (New York: Plenum Publishing, 1990); worldwide shipments are from Paul D. Maycock, "1993 World Module Ship-

ments," *Photovoltaic News*, February 1994; Victoria Griffith, "Twilight Hour," *Financial Times*, November 26, 1993.

9. Japanese sales from Maycock, "1993 World Module Shipments," op. cit. note 8.

10. Figure of 200,000 homes is from Neville Williams, Solar Electric Light Fund (SELF), Washington, D.C., private communication, January 14, 1994; Mark Hankins, *Solar Rural Electrification in the Developing World* (Washington, D.C.: SELF, 1993).

11. Richard Hansen, Enersol Associates, Inc., Somerville, Mass., private communication and printout, December 6, 1993.

12. Ibid.

13. Hankins, op. cit. note 10; Indonesian revolving credit from Netherlands Ministry of Foreign Affairs, *Sustainable Energy Economy*, Sector Policy Document of Development Cooperation 4 (Gravenhage: undated); Williams, op. cit. note 10; Richard D. Hansen and José G. Martin, "Photovoltaics for Rural Electrification in the Dominican Republic," *Natural Resources Forum*, Vol. 12, No. 2, 1988; solar lantern from "India's Photovoltaics Program Accelerates on All Fronts," *PV News*, July 1993.

14. Gerald Foley, *Electricity for Rural People* (London: Panos Institute, 1990); Hankins, op. cit. note 10.

15. Neville Williams, SELF, Washington, D.C., private communication, December 17, 1993; Charles Feinstein, World Bank, Washington, D.C., private communication and printout, December 20, 1993; Anil Cabraal, World Bank, Washington, D.C., private communication, December 8, 1993.

16. Electric Power Research Institute, *Technical Assessment Guide, Vol. 1: Electric Supply* (Palo Alto, Calif.: 1986); Mary Beth Regan, "The Sun Shines Brighter on Alternative Energy," *Business Week*, November 8, 1993; 500-watt system from Doug Pratt, Real Goods, Ukiah, Calif., private communication, March 1, 1994; Steven J. Strong, "An Overview of Worldwide Development Activity in Building-integrated Photovoltaics," Solar Design Associates, Inc., Harvard, Mass., undated.

17. Carlotta Collette, "Remote Possibilities," *Northwest Energy News*, July/August 1993.

18. Figure 8–3 is based on John Thornton, National Renewable Energy Laboratory, Golden, Colo., private communication and printout, April 1, 1994.

19. Zweibel, op. cit. note 8; "PV Efficiencies to Rise Sharply, Costs to Crumble by 2010: Maycock," *The Solar Letter*, July 23, 1993; Maycock, "1993 World Module Shipments," op. cit. note 8.

20. Figure of 33 percent is from Thornton, op. cit. note 18, and from Tom Surek, National Renewable Energy Laboratory,

Golden, Colo., private communication and printout, February 2, 1994; Taylor Moore, "High Hopes for High-Powered Solar," *EPRI Journal*, December 1992; Zweibel, op. cit. note 8; the manufacturer of the gallium arsenide cell claims to have achieved a 37-percent efficiency rating, but this came under conditions different from those standardly used by the National Renewable Energy Laboratory.

21. "Late News Development," *Photovoltaic Insider's Report*, April 3, 1991; Thomas E. Mallouk, "Bettering Nature's Solar Cells," *Nature*, October 24, 1991; Michael Grätzel, "Solar Cell Revolution," *The World and I*, February 1993.
22. Zweibel, op. cit. note 6.
23. Zweibel, op. cit. note 8.
24. Ibid.; Energy Conversion Devices, "Company History," Troy, Mich., undated; Subhendu Guha, United Solar Systems, Troy, Mich., private communication, December 10, 1993.
25. "New Silicon Cell Can Halve Cost of Solar Energy," *Wall Street Journal*, January 19, 1994; "USSC Panel Tests at Stable 10.3% [sic]; Production Line Running in Spring 1995," *The Solar Letter*, January 21, 1994.
26. Paul Maycock, "Boomer's Corner," *PV News*, May 1993; James H. Caldwell Jr., TGAL, Inc., Tracys Landing, Md., private communication, March 21, 1994.
27. Paul Maycock and Edward N. Stirewalt, *A Guide to the Photovoltaic Revolution* (Emmaus, Pa.: Rodale Press, 1985); Christopher C. Swan, *Sunlight: Energy, Economy & Photovoltaics* (San Francisco, Calif.: Sierra Club Books, 1986).
28. Maycock, "1993 World Module Shipments," op. cit. note 8.
29. Hankins, op. cit. note 10; Lovejoy, op. cit. note 2; Mark Hankins, "Home Systems are Fastest Growing Commercial PV Market in Africa," *PV News*, March 1992.
30. "Solec Expands Capacity, Reduces Cost, Patents 20% Process and Signs Joint Venture with India," *PV News*, October 1993; "India's Photovoltaics Program Accelerates on All Fronts," *PV News*, July 1993; Sherring Energy Associates, "N.R.E.L. India Initiative: Phase 1 Visit Report, January 22-February 11, 1994," for the National Renewable Energy Laboratory, Golden, Colo., undated; Energy Conversion Devices, op. cit. note 24.
31. Worldwatch estimate, based on United Nations, *1991 Energy Statistics Yearbook* (New York: 1993).
32. John Schaefer and Edgar DeMeo, Electric Power Research Institute, "An Update on U.S. Experiences with Photovoltaic Power Generation," Proceedings of the American Power Conference, April 23, 1990.
33. Carl Weinberg, Joseph J. Iannucci, and Melissa M. Reading,

"The Distributed Utility: Technology, Customer and Public Policy Changes Shaping the Electrical Utility of Tomorrow," Research and Development, Pacific Gas and Electric, San Ramon, Calif., December 1992.

34. Daniel S. Shugar, Howard J. Wenger, and Greg J. Ball, "Photovoltaic Grid Support: A New Screening Methodology," *Solar Today*, September/October 1993; national extrapolation is based on U.S. Department of Energy (DOE), Energy Information Administration (EIA), *Monthly Energy Review December 1993* (Washington, D.C.: U.S. Government Printing Office, 1993), and private communication from EIA, DOE, Washington, D.C., August 12, 1993.

35. Donald Osborn, Solar Program, Sacramento Municipal Utility District, Sacramento, Calif., private communication, February 22, 1994; Southern California Edison, "Southern California Edison and Texas Instruments Develop a Low Cost Solar Cell," Rosemead, Calif., undated.

36. Fred J. Sissine, "Renewable Energy: A New National Commitment?" Congressional Research Service, Washington, D.C., January 5, 1994; Ted Kennedy, Meridian Corporation, Alexandria, Va., private communication and printout, December 22, 1993.

37. DOE, "FY 1995 Budget Highlights," Washington, D.C., February 1994; Kennedy, op. cit. note 36.

38. "European Conference Hopes New Charter Will Boost Photovoltaics Along Swiss Model," *European Energy Report*, October 30, 1992; Strong, op. cit. note 16.

39. E.H. Lysen, Netherlands Agency for Energy and the Environment, "Solar Energy in the Netherlands," presented to the IEA-SHCP National Programs Workshops, Sydney, Australia, May 4, 1993; Strong, op. cit. note 16; "New Danish PV Manufacturer, New Government PV Program," *PV News*, October 1993; Koichi Yamanashi, executive managing director, New Energy Foundation, Tokyo, private communications, December 16, 1993, and February 17, 1994; "MITI to Subsidize Home Photovoltaics," *The Quad Report*, Washington, D.C., November 1993; "Sharp and Kansai Develop Solar Boosted Air Conditioner," *The Quad Report*, June 1993.

40. Utility PhotoVoltaic Group, "Electric Utilities Serving 40% of U.S. Consumers Propose $513 Million Program to Accelerate Use of Solar Photovoltaics," Washington, D.C., September 27, 1993.

41. Harvey Sachs, Center for Global Change, College Park, Md., private communication and printout, March 3, 1994; Harvey

M. Sachs, Frank Muller, and Alan S. Miller, "Getting Stakeholders Aboard the PV Train: Lessons from the 'Golden Carrot' Program and Other Adventures," in Jane Weissman, ed., *State Working Group Handbook* (Washington, D.C.: PV for Utilities, 1992).
42. Zweibel, op. cit. note 6.

CHAPTER 9. Plant Power

1. Ragnar Lundqvist, Bioflow/Ahlstrom, Varkaus, Finland, private communication, January 27, 1994; Ragnar Lundqvist, Martti Puhakka, and Krister Ståhl, "Pressurized CFB Biomass Gasification—The Bioflow Energy System," presented to the CFB4 Conference, Somerstet, Penn., August 1–5, 1993; Ragnar Lundqvist, "The IGCC Demonstration Plant at Värnamo," *Bioresource Technology*, Vol. 46, pp. 49–53, 1993.
2. Thomas B. Johansson et al., "Renewable Fuels and Electricity for a Growing World Economy: Defining and Achieving the Potential," in Thomas B. Johansson, Birgit Bodlund, and Robert H. Williams, eds., *Renewable Energy: Sources for Fuels and Electricity* (Washington, D.C.: Island Press, 1993).
3. United Nations (UN), *1991 Energy Statistics Yearbook* (New York: 1993); David Hall, King's College London, private communication, February 18, 1994, and printout, March 7, 1994; J.M.O. Scurlock and D.O. Hall, "The Contribution of Biomass to Global Energy Use," *Biomass*, No. 21, 1990; British Petroleum (BP), *BP Statistical Review of World Energy* (London: 1993); P.J. de Groot and D.O. Hall, "Biomass Energy: A New Perspective," prepared for African Energy Policy Research Network (AFREPREN), Third Workshop, University of Botswana, Gaborone, Botswana, December 1989; D.O. Hall, "Biomass Energy," *Energy Policy*, October 1991; Table 9–1 is based on Hall, private communication and printout, op. cit. this note, on UN, *1990 Energy Statistics Yearbook* (New York: 1992), and on BP, op. cit. this note.
4. U.S. Congress, Office of Technology Assessment (OTA), *Energy in Developing Countries* (Washington, D.C.: U.S. Government Printing Office (GPO), 1991); Kenneth Newcombe, "Economic Justification for Rural Afforestation: The Case of Ethiopia," in Gunter Schramm and Jeremy J. Warford, eds., *Environmental Management and Economic Development* (Baltimore, Md.: Johns Hopkins University Press, 1989); John Cogan, Energy Information Administration, U.S. Department of Energy (DOE), Washington, D.C., private communication,

February 24, 1994; David Pimentel, "Energy Security, Economics, and the Environment," *Journal of Agriculture and Environmental Ethics*, Vol. 4, No. 1, 1991.

5. Gautam S. Dutt and N.H. Ravindranath, "Bioenergy: Direct Applications in Cooking," in Johansson, Bodlund, and Williams, op. cit. note 2; health impact from World Bank, *World Development Report 1992* (New York: Oxford University Press, 1992); Kenya from Ogunlade Davidson and Stephen Karekezi, "A New, Environmentally-Sound Energy Strategy for the Development of Sub-Saharan Africa," AFREPREN, Nairobi, January 1992; China from Kirk Smith et al., "One Hundred Million Improved Cookstoves in China: How Was It Done?" *World Development*, Vol. 21, No. 6, 1993; Lundqvist, "The IGCC Demonstration Plant," op. cit. note 1.

6. Dean Mahin, "Industrial Energy and Electric Power from Wood Residues," Winrock International, Arlington, Va., June 1991; Worldwatch estimate based on D.O. Hall et al., "Biomass for Energy: Supply Prospects," in Johansson, Bodlund, and Williams, op. cit. note 2, and on Table 2–1, World Primary Energy Use.

7. John L. Preston et al., "Article: Energy Efficiency in the Manufacturing Sector," *Monthly Energy Review December 1992* (Washington, D.C.: GPO, 1992). The current U.S. biomass generating capacity excludes the roughly 1,800 megawatts of municipal solid waste incineration projects, and is found in Jane Turnbull, "Strategies for Achieving a Sustainable, Clean and Cost-effective Biomass Resource," Electric Power Research Institute (EPRI), Palo Alto, Calif., January 1993.

8. The Centre of Biomass Technology, "Straw for Energy Production: Technology—Environment—Economy," Aarhus, Denmark, 1992; European Parliament, Scientific and Technological Options Assessment, *Energy & Biomass: Country Profiles: Agricultural and Forestry Biomass Production—Operations Achieved* (Luxembourg: 1993).

9. "EU Aid Sought for Wood Plant," *European Energy Report*, January 7, 1994; William G. Mahoney, "Danes to Help German Firm with Straw-Burning Technology," *The Solar Letter*, August 20, 1993.

10. Robert H. Williams and Eric D. Larson, "Advanced Gasification-Based Biomass Power Generation," in Johansson, Bodlund, and Williams, op. cit. note 2.

11. Ibid.

12. A.E. Carpentieri et al., "Prospects for Sustainable, Utility-scale, Biomass-based Electricity Supply in Northeast Brazil," *Biomass*

and Bioenergy, Vol. 4, No. 3, 1993; cost estimates from Philip Elliott and Roger Booth, "Brazilian Biomass Power Demonstration Project," Special Project Brief, Shell International, London, September 1993.

13. See *International Cane Energy News*, Winrock International, Rosslyn, Va., July 1993 and January 1994, for information regarding developing-country projects.

14. U.K. from "Feature: Landfill Gas," *Warmer Bulletin*, August 1993; Germany from Andreas von Schoenberg, "Waste Management and Recycling in Germany," Andreas von Schoenberg Research & Consultancy, Dresden, Germany, 1993.

15. Based on a 100-year time horizon, from Intergovernmental Panel on Climate Change (IPCC), *Climate Change 1992: The IPCC Supplementary Report* (New York: Cambridge University Press, 1992); Peter Gleick et al., "Greenhouse-Gas Emissions from the Operation of Energy Facilities," prepared for the Independent Energy Producers Association, Sacramento, Calif., July 22, 1989.

16. Jeffrey Morris and Diana Canzoneri, "Comparative Lifecycle Energy Analysis: Theory and Practice," *Biocycle*, November 1992; Jeff Bailey, "Incinerator Becomes Financial Burden as Levels of Trash Trail Expectations," *Wall Street Journal*, November 12, 1993; Sandra Postel and John C. Ryan, "Reforming Forestry," in Lester R. Brown et al., *State of the World 1991* (New York: W.W. Norton & Company, 1991).

17. Patrick Knight, "Sugar Alcohol for $32 a Barrel," *Energy Economist*, August 1993; José Goldemberg, Lourival C. Monaco, and Isaias C. Macedo, "The Brazilian Fuel-Alcohol Program," in Johansson, Bodlund, and Williams, op. cit. note 2; José Roberto Moreira, executive director, Biomass Users Network, São Paulo, private communication and printout, February 24, 1994.

18. Crop conversion estimates are Worldwatch estimates, based on U.N. Food and Agriculture Organization, *FAO Production Yearbook* (Rome: 1993), on UN, op. cit. note 3, and on biomass-to-fuel conversion rates from Moreria, op. cit. note 17, and from Irshad Ahmed and David Morris, "Clearing the Air about Ethanol," *Carrying Capacity Network*, Vol 2., No. 3, 1992; for limited environmental benefit, see, for example, Organisation for Economic Co-operation and Development (OECD), International Energy Agency (IEA), "Biofuels," Paris, January 1994; Wim de Boo, "Environmental and Energy Aspects of Liquid Biofuels," Centrum voor Energiebesparing en Schone Technologie, Delft, the Netherlands, draft, February 1993; "Rape Oil–Based Diesel Fuel Found to be Environmentally Unsuit-

able," *Die Welt*, January 29, 1993, as reprinted in *JPRS Report: Environmental Issues*, March 29, 1993; "Finnish Report Questions Wisdom of Biofuels Aid as Helsinki Argues Case for Rural Gains," *Renewable Energy Report*, A European Energy Report Supplement, February 18, 1994; Pimentel, op. cit. note 4.

19. Ethanol prices are Worldwatch estimates, based on Paul Bergeron, National Renewable Energy Laboratory, Golden, Colo., private communication, April 5, 1994, and on a gasoline-to-ethanol conversion ratio from Goldemberg, Monaco, and Macedo, op. cit. note 17.

20. Jane Turnbull, "A 'How-to' Primer for Biomass Resource Development," EPRI, Palo Alto, Calif., December 1993; Leslie Lamarre, "Electricity from Whole Trees," *EPRI Journal*, January/February 1994.

21. Johansson, Bodlund, and Williams, op. cit. note 2; land set-aside in industrial countries is from U.S. Department of Agriculture, Washington, D.C., private communication, February 22, 1994, and from Lester R. Brown, "Facing Food Insecurity," in Lester R. Brown et al., *State of the World 1994* (New York: W.W. Norton & Company, 1994); energy production is Worldwatch estimate based on 10 tons dry material per hectare (which is an average based on 15 tons per hectare for cropland and 8 tons per hectare for environmentally fragile lands), based on Hall, private communication, op. cit. note 3, and on Turnbull, op. cit. note 7; industrial-country energy use is from OECD, IEA, *Energy Policies of IEA Countries: 1992 Review* (Paris: 1993).

22. R.L. Graham, L.L. Wright, and A.F. Turhollow, "The Potential for Short-Rotation Woody Crops to Reduce U.S. CO_2 Emissions," *Climatic Change*, November 1992; J. Pasztor and L. Kristoferson, "Bioenergy and the Environment: The Challenge," in Janos Pasztor and Lars A. Kristoferson, *Bioenergy and the Environment* (Boulder, Colo.: Westview Press, 1990); James H. Cook, Jan Beyea, and Kathleen H. Keeler, "Potential Impacts of Biomass Production in the United States on Biological Diversity," *Annual Review of Energy and the Environment* (Palo Alto, Calif.: Annual Reviews Inc., 1991); OTA, *Potential Environmental Impacts of Bioenergy Crop Production* (Washington, D.C.: GPO, 1993); EPRI, "EPRI and the National Audubon Society Establish the National Biofuels Roundtable," Palo Alto, Calif., December 1992.

23. Jan Beyea et al., "Toward Ecological Guidelines for Large-Scale Biomass Energy Development," report of a workshop for engineers, ecologists, and policymakers convened by the National Audubon Society and Princeton University, May 6, 1991.

24. Johansson, Bodlund, and Williams, op. cit. note 2.
25. Jodi L. Jacobson, *Gender Bias: Roadblock to Sustainable Development*, Worldwatch Paper 110 (Washington, D.C.: Worldwatch Institute, September 1992).
26. Gerald Leach and Robin Mearns, *Beyond the Woodfuel Crisis: People, Land and Trees in Africa* (London: Earthscan Publications Ltd., 1988); Zhao Tishun and Lu Qi, "Agroforestry on the North China Plain," *Agroforestry Today*, April–June 1993.
27. Gerald Leach, "Agroforestry and the Way Out for Africa," in Mohamed Suliman, *Greenhouse Effect and its Impact on Africa* (London: Institute for African Alternatives, 1990); David Brooks and Hartmut Krugmann, "Energy, Environment, and Development: Some Directions for Policy Research," *Energy Policy*, November 1990.
28. Keith Lee Kozloff and Roger C. Dower, *A New Power Base: Renewable Energy Policies for the Nineties and Beyond* (Washington, D.C.: World Resources Institute, 1993); ethanol example is from Dennis Anderson, *The Energy Industry and Global Warming: New Roles for International Aid* (London: Overseas Development Institute, 1992).
29. For a discussion of cropland availability, see Pierre Crosson, "Future Supplies of Land and Water for World Agriculture," Resources for the Future, Washington, D.C., December 1993; grain production from Brown, op. cit note 21; irrigation from Sandra Postel, "Irrigation Expansion Slowing," in Lester R. Brown, Hal Kane, and David Malin Roodman, *Vital Signs 1994* (New York: W.W. Norton & Company, 1994); UN, *Long-range World Population Projections: Two Centuries of Population Growth, 1950–2150* (New York: 1992).
30. Carl Hoffman, "Energy Futures," *Audubon*, September/October 1993; Cook, Beyea, and Keeler, op. cit. note 22.
31. Elliot and Booth, op. cit. note 12; Eric D. Larson, "Technology for Electricity and Fuels from Biomass," *Annual Review of Energy and the Environment* (Palo Alto, Calif.: Annual Reviews Inc., 1993); Sandy Thomas, Office of Senator Tom Harkin, U.S. Senate, Washington, D.C., private communication, April 29, 1994.
32. David Morris and Irshad Ahmed, *The Carbohydrate Economy: Making Chemicals and Industrial Materials from Plant Matter* (Washington, D.C.: Institute for Local Self-Reliance, 1992).
33. UN, op. cit note 3; BP, op. cit. note 3.
34. Joseph Kahn, "Damned Yangtze: Despite Vast Obstacles, Chinese Move to Tap Power of Historic River," *Wall Street Journal*, April 18, 1994; World Wildlife Fund–India, *Dams on the Nar-*

mada: A People's View (New Delhi: 1986); future supply figures are authors' estimates; for further scenario details, see Chapter 13.

35. Ronald DiPippo, "Geothermal Energy," *Energy Policy*, October 1991.
36. World geothermal capacity is a Worldwatch estimate based on UN, op. cit. note 3, and on Gerald Huttrer, Geothermal Management Company, Inc., Frisco, Colo., private communication, February 10, 1994; Civis G. Palmerini, "Geothermal Energy," in Johansson, Bodlund, and Williams, op. cit. note 2; Edwin Karmiol, "Japan Awakening to Potential of Domestic Geothermal Energy," *The Solar Letter*, November 13, 1992.
37. D.H. Freeston, "Direct Uses of Geothermal Energy in 1990," *Geothermal Resources Council Bulletin*, July/August 1990.
38. DOE, "U.S. Geothermal Energy R&D Program Multi-Year Plan, 1988–1992," Washington, D.C., 1988.
39. Projections are authors' estimates; for a discussion of Worldwatch scenario, see Chapter 13.

CHAPTER 10. Reinventing Transportation

1. Personal communications with various people involved in the ZEV process.
2. Stacy C. Davis and Melissa D. Morris, *Transportation Energy Data Book: Edition 12* (Oak Ridge, Tenn.: Oak Ridge National Laboratory (ORNL), 1992); "EV Technology Moves Forward as Siege of ZEV Mandate Continues," *Green Car Journal*, April 1994.
3. "Northeast Votes to Adopt Calif. Standards," *Green Car Journal*, February 1994.
4. Details and supporting evidence are presented later in the chapter.
5. Figure 10–1 is from American Automobile Manufacturers Association, *World Motor Vehicle Data 1993* (Detroit, Mich.: 1993).
6. Eric Toler, Rocky Mountain Institute, Snowmass, Colo., private communication, March 30, 1994.
7. Urs Muntwyler, "The Promotion Program for Lightweight Electric Vehicles in Switzerland," IngenieurbÜro Muntwyler, Zollikofen, Switzerland, paper presented to the Swiss Federal Office of Energy Economy, undated.
8. Ibid.
9. Christopher Flavin and Michael Renner, "Solar Cars Race Across Australia," *World Watch*, March/April 1988.
10. Ibid.

11. Paul MacCready, "Electric Vehicles: Great Potential, Great Confusion," AeroVironment Inc., Monrovia, Calif., unpublished, March 22, 1993.
12. American Automobile Manufacturers Association, *Facts and Figures '93* (Detroit, Mich.: 1993).
13. Nicholas Lenssen and John Young, "Filling Up in the Future," *World Watch*, May/June 1990.
14. Sam H. Schurr et al., *Electricity in the American Economy* (New York: Greenwood Press, 1960).
15. Amory B. Lovins, John W. Barnett, and L. Hunter Lovins, "Supercars: The Coming Light-Vehicle Revolution," Rocky Mountain Institute, Snowmass, Colo., 1993.
16. Ibid.; "EC Seen Approving French/German Plan For Recycling of Auto, Electronic Goods," *International Environment Reporter*, January 13, 1993; "Daimler-Benz, Mitsubishi Agree On Two Automobile Recycling Projects," *International Environment Reporter*, December 15, 1993.
17. Switzerland from conversation with Prof. Walz, Federal University of Technology, Zurich, concerning crash tests carried out at Wildhaus on August 18, 1992.
18. Cliff Gromer, "New Age of the Electric Car," *Popular Mechanics*, February 1994; Dan McCosh, "We Drive the World's Best Electric Car," *Popular Science*, January 1994.
19. McCosh, op. cit. note 18; California Air Resources Board (CARB), "Draft Discussion Paper for the Low-Emission Vehicle Workshop," Sacramento, Calif., March 25, 1994.
20. Paul B. MacCready, "Lightweight Construction and Aerodynamics," paper presented at the Conference on Electric Hybrids, Zurich, Switzerland, September 20, 1993; Amory B. Lovins, "Rethinking Efficient Cars," Rocky Mountain Institute, Snowmass, Colo., 1993.
21. Lovins, op. cit. note 20.
22. Lovins, Barnett, and Lovins, op. cit. note 15.
23. Hybrid vehicle developments from *Green Car Journal*, various issues; John S. Reuyl et al., "A Comparison of the Hybrid Electric Vehicle with the Pairing of a Battery-Only Electric Vehicle and Gasoline Vehicle," Nevcor, Inc., Stanford, Calif., unpublished, November 15, 1993; U.S. Department of Energy (DOE), "Hybrid Vehicle Program Plan," Washington, D.C., June 1993; Table 10–1 is based on Bill King, Ford Motor Company, Washington, D.C., private communication and printout, March 30, 1994, on Chris Marshall, American Honda, Torrance, Calif., private communication and printout, April 5, 1994, on CARB, "ARB Certifies First Ultra Low Emission Ve-

hicle," press release, Sacramento, Calif., January 7, 1994, on
"Volvo ECC: A Volvo Experimental Concept Car," product
literature, Volvo Cars of North America, Rockleigh, N.J., on
Energy International, Inc., "Comparison of Emissions for Low
Emissions Gasoline and Alternative Fuel Vehicles: Discussion
Draft," Gas Research Institute, Chicago, 1993, on Gromer, op.
cit. note 18, on CARB, "Zero-Emission Vehicle Update," Sac-
ramento, Calif., March 1994, and on Bevilacqua Knight, Inc. et
al., "1992 Electric Vehicle Technology and Emissions Update,"
Sacramento, Calif., 1992.

24. Muntwyler, op. cit. note 7.

25. Horlacher AG, promotional materials, Möhlin, Switzerland,
January 1993.

26. Urs Muntwyler, Zollikofen, Switzerland, private communication, February 4, 1993.

27. *Green Car Journal*, op. cit. note 23.

28. Various vehicle developments are described in ibid.; Paul B.
MacCready, "Jumpstart: The Sub-Car Electric Vehicle Oppor-
tunity," paper presented at AeroVironment Inc. Workshop,
Monrovia, Calif., January 28, 1993.

29. Worldwatch estimate based on specifications in Lovins, Barnett,
and Lovins, op. cit. note 15.

30. Worldwatch estimate, assuming a fuel efficiency equivalent to
2.3 liters per 100 kilometers (100 miles per gallon) for natural
gas hybrids, and the Impact's fuel effiency of 0.35 kilowatt-
hours per kilometer for electrics, and based on DOE, Energy
Information Administration (EIA), *Annual Energy Review 1992*
(Washington, D.C.: U.S. Government Printing Office (GPO),
1993), and on U.S. Department of Transportation, Federal
Highway Administration, *Highway Statistics 1991* (Washington,
D.C.: 1992).

31. "Japan Advancing Electric and Low-Emission Vehicle Plans,"
Green Car Journal, January 1994.

32. Robert H. Williams, "The Clean Machine," *Technology Review*,
April 1994; Philip H. Abelson, "Applications of Fuel Cells"
(editorial), *Science*, June 22, 1990.

33. "The Push for Advanced Batteries," *EPRI Journal*, April/May
1991.

34. W.H. DeLuca et al., "Results of Advanced Battery Technology
Evaluations for Electric Vehicle Applications," Society of Automotive Engineers, Warrendale, Pa., 1992; CARB, "Zero-Emission Vehicle Update," op. cit. note 23.

35. Richard F. Post, "Flywheel Energy Storage," *Scientific American*, September 1973; John V. Coyner, "Flywheel Energy Stor-

age and Power Electronics Program," ORNL, Oak Ridge, Tenn., 1993; Abacus Technology Corporation, "Technology Assessments of Advanced Energy Storage Systems for Electric and Hybrid Vehicles," prepared for U.S. Department of Energy, April 30, 1993; Lawrence Livermore projections from Richard F. Post, Lawrence Livermore National Laboratory, Livermore, Calif., private communication, July 15, 1993.

36. Post, private communication, op. cit. note 35.
37. Lovins, Barnett, and Lovins, op. cit. note 15.
38. Donald W. Nauss and Michael Parrish, "Big 3 Try to Put Brakes on Push for Electric Cars," *Los Angeles Times*, October 23, 1993.
39. *Green Car Journal*, op. cit. note 23.
40. Matthew L. Wald, "In Quest for Electric Cars, He Adds the Power of Faith," *New York Times*, March 6, 1994; "Electricar Buys Synergy EV Group," *Green Car Journal*, December 1993; "Mercedes to Produce Swatch EV," *Green Car Journal*, April 1994.
41. Calstart, *Summary of Advanced Transportation Technologies* (Burbank, Calif.: 1993).
42. Matthew L. Wald, "Expecting a Fizzle, G.M. Puts Electric Car to Test," *New York Times*, January 28, 1994.
43. Marcia D. Lowe, *Alternatives to the Automobile: Transport for Livable Cities*, Worldwatch Paper 98 (Washington, D.C.: Worldwatch Institute, October 1990).
44. Marcia D. Lowe, *Back on Track: The Global Rail Revival*, Worldwatch Paper 118 (Washington, D.C.: Worldwatch Institute, April 1994).
45. Ibid.
46. Marcia D. Lowe, *The Bicycle: Vehicle for a Small Planet*, Worldwatch Paper 90 (Washington, D.C.: Worldwatch Institute, September 1989).
47. Ibid.
48. The wholesale price of gasoline in the United States was 17¢ per liter (63¢ per gallon) in 1993, according to DOE, EIA, *Monthly Energy Review March 1994* (Washington, D.C.: GPO, 1994), compared with an estimated postwar low of 15¢ per liter (58¢ per gallon) in 1972 (1993 dollars), based on ibid., and on American Petroleum Institute, *Basic Petroleum Data Book* (Washington, D.C.: 1985).
49. Per Kågeson, *External Costs of Air Pollution: The Case of European Transport* (Brussels: European Federation for Transport and Environment, 1992); International Union of Railways, *Internalization of External Effects in Transportation* (Brussels: 1992).

50. John DeCicco et al., "Feebates for Fuel Economy: Market Incentives for Encouraging Production and Sales of Efficient Vehicles," (Washington, D.C.: American Council for an Energy-Efficient Economy, 1993); *Green Car Journal*, op. cit. note 23.

51. CARB, "Zero-Emission Vehicle Update," op. cit. note 23; Alan Riding, "Swiss Give New Meaning to the Word Roadblock," *New York Times*, February 28, 1994.

52. Diane Wittenberg, Southern California Edison, Rosemead, Calif., private communication, June 1, 1994.

CHAPTER 11. Building for the Future

1. ING Bank example is from Internationale Nederlanden Bank, "Building with a Difference: ING Bank Head Office," Amsterdam, undated, from William Browning, "NMB Bank Headquarters: The Impressive Performance of A Green Building," *Urban Land*, June 1992, from Bill Holdsworth, "Organic Services," *Building Services*, March 1989, and from Rob Vonk, ING Bank, Amsterdam, private communication, March 25, 1994.

2. Browning, op. cit. note 1; Holdsworth, op. cit. note 1.

3. William Browning, Rocky Mountain Institute, Snowmass, Colo., private communication, March 28, 1994; Vonk, op. cit. note 1.

4. McDonough quoted in Jean Gorman, "Technology in the Age of Interdependence," *Interiors*, March 1993.

5. Global building energy consumption is a Worldwatch Institute estimate based on Organisation for Economic Co-operation and Development (OECD), International Energy Agency (IEA), *Energy Statistics and Balances of Non-OECD Countries 1990–1991* (Paris: 1993), and on OECD, IEA, *World Energy Statistics and Balances 1971–1987* (Paris: 1989).

6. Residential energy use statistics are based on energy use per degree day per square meter of home area, and are from Lee Schipper and Stephen Meyers, *Energy Efficiency and Human Activity: Past Trends, Future Prospects* (New York: Cambridge University Press, 1992), and from L. Schipper and C. Sheinbaum, "Recent Trends in Household Energy Use Efficiency in OECD Countries: Stagnation or Improvement," in *Proceedings of 1994 ACEEE Workshop on Energy Efficiency in Buidlings* (Washington, D.C.: American Council for an Energy-Efficient Economy, in press).

7. Jayant Sathaye and Ashok Gadgil, "Aggressive Cost-Effective Electricity Conservation," *Energy Policy*, February 1992.

8. "Facts on ACT²," Pacific Gas and Electric, San Ramon, Calif.,

October 1990; Grant Brohard, project technology engineer, ACT², Pacific Gas and Electric, San Ramon, Calif., private communication, July 7, 1993; Carl Weinberg, Weinberg and Associates, Walnut Creek, Calif., private communication, November 4, 1993.

9. Schipper and Meyers, op. cit. note 6; S. Meyers, L. Schipper, and J. Salay, "Energy Use in Poland, An International Comparison," *Energy, The International Journal*, in press; L. Schipper et al., "The Structure and Efficiency of Energy Use in Estonia," Lawrence Berkeley Laboratory, Berkeley, Calif., forthcoming; U.S. Congress, Office of Technology Assessment, *Energy Efficiency Technologies for Central and Eastern Europe* (Washington, D.C.: U.S. Government Printing Office, 1993).

10. Sathaye and Gadgil, op. cit. note 7.

11. Alan Thein Durning, "Home and Hearth" (unpublished manuscript), Northwest Environment Watch, Seattle, Wash., undated; Tim Mayo, Department of Natural Resources, Ottawa, "Canada's Advanced Houses Program," Green Buildings Conference, Gaithersburg, Md., February 16–17, 1994.

12. Carl Weinberg, Weinberg and Associates, Walnut Creek, Calif., private communication, April 23, 1993; Amory Lovins, "Hot-Climate House Predicted to Need No Air Conditioner, Cost Less to Build," Tech Memo, E SOURCE, Inc., Boulder, Colo., November 1993.

13. Steven Ternoey et al., *The Design of Energy-Responsive Commercial Buildings* (New York: John Wiley & Sons, 1985); Weinberg, op. cit note 12.

14. Karen L. George, "Highly Reflective Roof Surfaces Reduce Cooling Energy Use and Peak Demand," Tech Update, E SOURCE, Inc., Boulder, Colo., December 1993; Lawrence Berkeley Laboratory and Sacramento Municipal Utility District, "Peak Power and Cooling Energy Savings of Shade Trees and White Surfaces: Year 2," Lawrence Berkeley Laboratory, Berkeley, Calif., April 27, 1993.

15. Steven Ternoey, LightForms, Boulder, Colo., private communication, March 29, 1994; Alicia Ravetto, "Daylighting Schools in North Carolina," *Solar Today*, March/April 1994.

16. Neuffer from Donald Aitken and Paul Bony, "Passive Solar Production Housing and the Utilities," *Solar Today*, March/April 1993; Donald Aitken, Union of Concerned Scientists, Woodside, Calif., private communication, April 5, 1994; Doug Balcomb, National Renewable Energy Laboratory (NREL), Golden, Colo., private communication and printout, June 2, 1993.

17. Mayo, op. cit. note 11; "New Canadian Program Soars Twice as Far as R-2000 Homes," *The Solar Letter*, March 4, 1994.
18. Steven J. Strong, with William G. Scheller, *The Solar Electric House: Energy for the Environmentally-Responsive, Energy-Independent Home* (Still River, Mass.: Sustainability Press, 1993); Steven Strong, Solar Design Associates, Harvard, Mass., private communication, March 30, 1994.
19. Joachim Brenemann, "Energy Active Façades: Technology and Possibilities of Photovoltaic Integration into Buildings," Flachglas Solartechnik, Köln, Germany, undated; Flagsol, Flachglas Solartechnik Gmbh, "Reference List," Köln, Germany, January 14, 1994; "Siemens Solar and Corning to Cooperate on CIS Development," *PV News*, March 1994; Steven Strong, "An Overview of Worldwide Development Activity in Building-Integrated Photovoltaics," Solar Design Associates, Harvard, Mass., undated.
20. J.P. Louineau et al., IT Power Ltd., Eversley, Hants, U.K., "Photovoltaic Cladding Systems for Commercial Buildings in UK," undated; R. Hill, N.M. Pearsall, and P. Claiden, *The Potential Generating Capacity of PV-Clad Buildings in the UK*, Vol. 1 (London: Department for Trade and Industry, 1992); Germany from W.H. Bloss et al., "Grid-Connected Solar Houses," in *Proceedings of the 10th EC Photovoltaics Solar Energy Conference* (Dordrecht: Kluwer Academic Publishing, 1991); OECD, IEA, *Energy Policies of IEA Countries: 1991 Review* (Paris: 1992); examples assume a 25-percent capacity factor for photovoltaics.
21. Amory B. Lovins, "Energy-Efficient Buildings: Institutional Barriers and Opportunities," Strategic Issues Paper, E SOURCE, Inc., Boulder, Colo, January 1994.
22. Bigelow from ibid. and from Cynthia Cocchi, The Bigelow Group, Inc., Palatine, Ill., private communication, March 29, 1994; Mark Congling, "A Solar Community: Eldorado at Santa Fe," *Solar Today*, November/December 1991; Kim Hamilton, "Village Homes," *In Context*, Late Spring 1993.
23. Erik Toxværd Nielsen, Toftegård, Herlev, Denmark, private communication, April 20, 1994.
24. Lovins, op. cit. note 21.
25. Building Research Establishment, "BREEAM/Existing Offices," Version 4/93, Garston, Watford, U.K.; Michael Scholand, "Buildings for the Future," *World Watch*, November/December 1993.
26. Raymond Cole, University of British Columbia, "Building Environmental Performance Assessment Criteria (BEPAC)," presented to the NIST/Green Building Council Conference, Gai-

thersburg, Md., February 16–17, 1994; U.S. Green Building Council, "Building Rating System," Staff Draft Issue Paper, for discussion at March 7, 1994, executive, public, and task force briefings; City of Austin, Environmental and Conservation Services Department, "Green Building Guide: A Sustainable Approach," Austin, Tex., September 1992; Lawrence Doxsey, Green Builder Program, Austin, Tex., private communication and printout, April 1, 1994; U.S. Environmental Protection Agency (EPA), Air and Radiation, "Introducing. . .The Energy Star Buildings Program," Washington, D.C., November 1993; Chris O'Brien, EPA, Washington, D.C., private communication, March 9, 1994.

27. Amory Lovins, "Designing Buildings for Greater Profit," presentation at the National Association of Homebuilders, Washington, D.C., March 2, 1994.

28. Lovins, op. cit. note 21.

29. Lovins, op. cit. note 27; see also Ternoey, op. cit. note 13.

30. Tim Lougheed, "R-2000," *Canadian Consumer*, May/June 1992.

31. Sweden from Lee Schipper, Stephen Meyers, and Henry Kelly, *Coming in From the Cold: Energy-Wise Housing in Sweden* (Cabin John, Md.: Seven Locks Press, 1985), and from Lee Schipper, Lawrence Berkeley Laboratory, Berkeley, Calif., private communication, April 1, 1994; Barbara Farhar and Jan Eckert, "Energy-Efficient Mortgages and Home Energy Rating Systems: A Report on the Nation's Progress," NREL, Golden, Colo., September 1993; Barbara Farhar, NREL, Washington, D.C., private communication, April 1, 1994.

32. "R-2000 Reaches New Milestone," *R-2000 News Communiqué*, Energy, Mines, and Resources Canada, Ottawa, February 1993; David Grafstein, Ontario Hydro, Toronto, private communication, April 6, 1994; Gary Sharp, R-2000 Program, Energy, Mines, and Resources Canada, Ottawa, private communication, March 31, 1994.

33. President's Commission on Environmental Quality and Alliance to Save Energy (ASE), "Guidelines for Energy Efficient Commercial Leasing Practices," ASE, Washington, D.C., October 1992; Lovins, op. cit. note 27; Lovins, op. cit. note 21.

34. This section relies heavily on Nicholas Lenssen and Marcia D. Lowe, "From Light Bulbs to Light Rails," *Design Spirit*, Winter 1991, and on Marcia D. Lowe, *Alternatives to the Automobile: Transport for Livable Cities*, Worldwatch Paper 98 (Washington, D.C.: Worldwatch Institute, October 1990); building and transportation energy use is based on OECD, *Energy Statistics 1990–*

1991, op. cit. note 5, and on OECD, *World Energy Statistics 1971–1987*, op. cit. note 5.

35. Susan Owens, University of Cambridge, "Cities and Sustainability: The Energy Dimension," presented to the Innovative Housing 93 Conference, Vancouver, June 1993.

36. Real Estate Research Corporation, "The Costs of Sprawl: Environmental and Economic Cost of Alternative Residential Patterns at the Fringe," prepared for EPA, Washington, D.C., 1974; Susan E. Owens, "Land Use Planning for Energy Efficiency," in J.B. Cullingworth, ed., *Energy, Land, and Public Policy* (New Brunswick, N.J.: Transaction Publishers, 1990).

37. Peter Newman and Jeffrey Kenworthy, *Cities and Automobile Dependence: An International Sourcebook* (Aldershot, U.K.: Gower, 1989).

38. Robert Cervero, "Transportation and Urban Development: Perspectives for the Nineties," Working Paper 470, University of California, Berkeley, August 1987.

39. Patrick Michell, Southern California Association of Governments, Los Angeles, private communication, March 31, 1994.

CHAPTER 12. Reshaping the Power Industry

1. Peter Asmus, "Saving Energy Becomes Company Policy," *Amicus Journal*, Winter 1993.

2. Energy Policy Project of the Ford Foundation, *A Time to Choose: America's Energy Future* (Cambridge, Mass: Ballinger, 1974).

3. Andy Asher, Sacramento Municipal Utility District (SMUD), Sacramento, Calif., private communication, September 17, 1993; SMUD, *1992 Annual Report: Working for Sacramento* (Sacramento, Calif.: undated); transcript from "Living on Earth," National Public Radio, August 27, 1993; SMUD, "Official Statement Relating to Sacramento Municipal Utility District $498,410,000 Electric Revenue Refunding Bonds, 1993 Series D, $75,000,000 Electric Revenue Bonds, 1993 Series E," Sacramento, Calif., April 15, 1993; Freeman quoted in Matthew L. Wald, "New Enterprises in a Nonnuclear Age," *In Business*, November/December 1993.

4. Worldwatch estimate based on Organisation for Economic Co-operation and Development (OECD), International Energy Agency (IEA), *Energy Statistics and Balances of Non-OECD Countries, 1990–1991* (Paris: 1993), and on D.O. Hall, King's College London, private communication, February 18, 1994, and printout, March 7, 1994; utility industry size is Worldwatch estimate based on United Nations (UN), *1990 Energy Statistics*

Yearbook (New York: 1992), and on Edison Electric Institute (EEI), *Statistical Yearbook of the Electric Utility Industry 1991* (Washington, D.C.: 1992); world auto industry size is a Worldwatch estimate based on U.S. Department of Commerce, Bureau of the Census, *Statistical Abstract of the United States 1992* (Washington, D.C.: 1992), on American Automobile Manufacturers Association, *World Motor Vehicle Data 1993* (Detroit, Mich.: 1993), and on Motor Vehicles Manufacturers Association of the United States, Inc., *Facts & Figures '90* (Detroit, Mich.: 1990).

5. Richard Munson, *The Power Makers* (Emmaus, Pa.: Rodale Press, 1985).

6. Electricity costs are from Richard F. Hirsh, *Technology and Transformation in the American Electric Utility Industry* (New York: Cambridge University Press, 1989), from EEI, "EEI Pocketbook of Electric Utility Industry Statistics," Washington, D.C., 1992, and from EEI, *Historical Statistics of the Electric Utility Industry Through 1970* (New York: 1973); Figure 12–1 is based on Hirsh, op. cit. in this note, U.S. Department of Commerce, Bureau of the Census, *Historical Statistics of the United States, Colonial Times to 1970,* Part 2 (Washington, D.C.: 1975), and on U.S. Department of Energy (DOE), Energy Information Administration (EIA), *Annual Energy Renew 1992* (Washington, D.C.: U.S. Government Printing Office (GPO), 1993).

7. Irvin C. Bupp and Jean-Claude Derian, *Light Water: How the Nuclear Dream Dissolved* (New York: Basic Books, Inc., 1978); International Atomic Energy Agency, *Nuclear Power Reactors in the World* (Vienna: 1993); Greenpeace International, WISE-Paris, and Worldwatch Institute, *World Nuclear Industry Status Report: 1992* (London: Greenpeace International, 1992).

8. Figure 12–2 is based on UN, *World Energy Supplies 1950–1974* (New York: 1976), on OECD, IEA, *Energy Balances of OECD Countries* (Paris: various years), on International Monetary Fund, *World Economic Outlook May 1993* (Washington, D.C.: 1993), on Robert Summers and Alan Heston, "The Penn World Table (Mark 5): An Expanded Set of International Comparisons, 1950–1988," *Quarterly Journal of Economics,* May 1991 (based on purchasing power parity), and on Yoko Takahashi, IEA, OECD, Paris, private communication and printout, April 29, 1994; French debt from "Export Surge Propels EdF from Strength to Strength," *European Energy Report,* February 18, 1994; Richard Rudolph and Scott Ridley, *Power Struggle: The Hundred-Year War Over Electricity* (New York: Harper & Row, 1986).

9. Table 12–1 is based on OECD, IEA, *World Energy Outlook* (Paris: 1993), and on OECD, op. cit. note 4; Barbara J. Cummings, *Dam the Rivers, Damn the People* (London: Earthscan Publications Ltd., 1990).

10. World Bank, *The World Bank's Role in the Electric Power Sector* (Washington, D.C.: 1993).

11. Ibid.

12. "Germany," Country Profiles, *European Energy Report*, May 1992; Jennifer S. Gitlitz, "The Relationship Between Primary Aluminum Production and the Damming of World Rivers," Energy and Resources Group, University of California, Berkeley, Calif., July 21, 1993; François Nectoux, *Crisis in the French Nuclear Industry* (Amsterdam: Greenpeace International, 1991).

13. Christopher Flavin, *Electricity's Future: The Shift to Efficiency and Small-Scale Power*, Worldwatch Paper 61 (Washington, D.C.: Worldwatch Institute, November 1984); California renewables capacity is from Jan Hamrin and Nancy Rader, *Investing in the Future: A Regulator's Guide to Renewables* (Washington, D.C.: National Association of Regulatory Utility Commissioners, 1993); percent of electricity supply is from Karen Griffin, California Energy Commission, private communication and printout, April 26, 1994.

14. Asea Brown Boveri and Texaco from Lester P. Silverman, McKinsey & Co., Inc., Washington, D.C., private communication, October 13, 1993; Figure 12–3 is based on DOE, EIA, *Annual Electric Generator Report* (electronic database) (Washington, D.C.: 1993), and on EEI, *Capacity and Generation of Non-Utility Sources of Energy* (Washington, D.C.: various years), with 1992 data from Utility Data Institute, *UDI Directory of Selected U.S. Cogeneration, Small Power and Industrial Power Plants* (Washington, D.C.: 1993). Independent producer data through 1991 is the amount of capacity placed in operation each year and still operating in 1991, thus the data underestimate the total amount of new capacity that was installed annually by independent producers.

15. Ministry of Energy, Danish Energy Agency, *Energy Efficiency in Denmark* (Copenhagen: 1992); Sara Knight, "German Survey," *Windpower Monthly*, March 1993; "Germany," op. cit. note 12; Steven Strong, "An Overview of Worldwide Development Activity in Photovoltaics," Solar Design Associates, Harvard, Mass., 1993; "Netherlands," Country Profiles, *European Energy Report*, March 1992; Matthew Parris, "The End of the Nuclear Affair," (London) *Times*, November 10, 1989; Andrew Holmes, "Electricity in Europe: Power and Profit," *Financial Times Man-*

agement Report, London, 1990; Sandy Hendry, "US, Europe Firms Look to Spark China Power Ventures," *Journal of Commerce,* June 2, 1993; U.S. Agency for International Development, Office of Energy and Infrastructure, "Country Profiles on India, Indonesia, and Pakistan," *Private Power Reporter,* March 1993; Andy Pasztor, "Power Plants in Mexico Cast Pall Over Nafta," *Wall Street Journal,* September 9, 1993.

16. DOE, op. cit. note 14; EEI, op. cit. note 14.
17. Table 12–2 is based on DOE, EIA, *Historical Plant Cost and Annual Production Expenses for Selected Electric Plants 1987* (Washington, D.C.: GPO, 1989), on DOE, EIA, *Electric Plant Cost and Power Production Expenses* (Washington, D.C.: GPO, various years), on Charles Komanoff, Komanoff Energy Associates, New York, private communication and printout, February 9, 1989, and on materials cited in Chapters 5, 6, and 7.
18. U.S. Congress, "National Energy Policy Act of 1992," PL 102–486, Washington, D.C., October 24, 1992.
19. David Roe, *Dynamos and Virgins* (New York: Random House, 1984); Hirsh, op. cit note 6.
20. OECD, IEA, *Energy Policies of IEA Countries, 1991 Review* (Paris: 1992).
21. Chris J. Calwell and Ralph C. Cavanagh, "The Decline of Conservation at California Utilities: Causes, Costs and Remedies," Natural Resources Defense Council (NRDC), San Francisco, Calif., July 1989; David Moskovitz, "Profits and Progress Through Least-Cost Planning," National Association of Regulatory Utility Commissioners, Washington, D.C., November 1989.
22. Rowe quote is found in EEI, *Washington Letter,* Washington, D.C., September 15, 1989; Moskovitz, op. cit. note 21; Stephen Wiel, "Making Utility Efficiency Profitable," *Public Utilities Fortnightly,* July 1989; Michael Smith, "An Island's Experience," *Financial Times,* September 8, 1993.
23. Cynthia Mitchell, consulting economist, Reno, Nev., private communication, June 24, 1993; Cynthia Mitchell, "Integrated Resource Planning Survey: Where the States Stand," *The Electricity Journal,* May 1992.
24. Linda Bromley, EIA, DOE, Washington, D.C., private communication, December 13, 1993; Electric Power Research Institute estimates cited in Amory Lovins, "Apples, Oranges, and Horned Toads: Is the Joskow & Marron Critique of Electric Efficiency Costs Valid?" *The Electricity Journal,* May 1994; Paul L. Joskow and Donald B. Marron, "What Does Utility-Subsidized Energy Efficiency Really Cost?" *Science,* April 16, 1993;

Amory B. Lovins, "The Cost of Energy Efficiency" (letter to the editor), *Science*, August 20, 1993.

25. John Fox, Ontario Hydro, Toronto, private communication, September 29, 1993; Katrina van Bylandt, Power Smart Inc., Vancouver, B.C., private communication, January 28, 1993; Evan Mills, "Efficient Lighting Programs in Europe: Cost Effectiveness, Consumer Response, and Market Dynamics," *Energy—The International Journal*, Vol. 18, No. 2, 1993; Evan Mills, Lawrence Berkeley Laboratory, Berkeley, Calif., private communication, June 15, 1993; Wim Sliepenbeek, "Massive Programs Get the Dutch Market Moving," *IAEEL Newsletter* (International Association for Energy-Efficient Lighting, Stockholm), No. 1, 1993; Uwe Leprich, "German Giant Explores Its Demand-Side Resources," *IAEEL Newsletter*, No. 2, 1992; Reinhard Loske, Wuppertal Institute for Climate, Environment and Energy, Wuppertal, Germany, private communication, September 24, 1993.

26. Thailand from Peter du Pont, Terry Kraft-Oliver, and Peter Rumsey, International Institute for Energy Conservation (IIEC), Washington, D.C., private communication, June 23, 1993; Howard S. Geller and José Roberto Moreira, "Brazil Encourages Electricity Savings," *Forum for Applied Research and Public Policy*, University of Tennessee, Fall 1993; Mark D. Levine, Feng Liu, and Jonathan E. Sinton, "China's Energy System: Historical Evolution, Current Issues, and Prospects," in *Annual Review of Energy and the Environment* (Palo Alto, Calif.: Annual Reviews Inc., 1992); Ignacio Rodríguez and David Wolcott, "Growth Through Conservation: DSM in Mexico," *Public Utilities Fortnightly*, August 1, 1993; Lawrence Berkeley Laboratory estimate is from Mark D. Levine et al., *Energy Efficiency, Developing Nations, and Eastern Europe*, A Report to the U.S. Working Group on Global Energy Efficiency (Washington, D.C.: IIEC, 1991), and from Charles Campbell, Lawrence Berkeley Laboratory, Berkeley, Calif, private communication and printout, June 19, 1992; Howard Geller, *Efficient Electricity Use: A Development Strategy for Brazil* (Washington, D.C.: American Council for an Energy-Efficient Economy, 1991); Lee Schipper and Eric Martinot, "Decline and Rebirth: Energy Demand in the Former Soviet Union" (draft), Paper II: Towards Efficiency in 2010, Lawrence Berkeley Laboratory, Berkeley, Calif., September 1992; David Wolcott, Jaroslaw Dybowski, and Ewaryst Hille, "Implementing Demand-Side Management Through Integrated Resource Planning in Poland," presented at the European Council for an Energy-Efficient Economy Summer Study, Rungstedgaard, Denmark, May 1993.

27. Electric Power Research Institute, *Drivers of Electricity Growth and the Role of Utility Demand-Side Management* (Palo Alto, Calif.: 1993); "Holland Turns the Tide," *IAEEL Newsletter*, No. 1, 1992; Greg M. Rueger, senior vice president and general manager, Pacific Gas and Electric Company (PG&E), Testimony before the Subcommittee on Energy and Power, Committee on Commerce and Energy, U.S. House of Representatives, Washington, D.C., March 7, 1991; Figure 12–4 is based on U.S. Department of Commerce, Bureau of the Census, *Statistical Abstract of the United States* (Washington, D.C.: various years), on State of California, Department of Finance, "Population Estimates for California Cities and Counties," Report 92 E-2, Sacramento, Calif., 1993, on DOE, EIA, op. cit. note 6, on DOE, EIA, *State Energy Data Report 1991: Consumption Estimates* (Washington, D.C.: various years), on Population Reference Bureau, *1992 World Population Data Sheet* (Washington, D.C.: 1992), and on Daniel Nix, California Energy Commission, Sacramento, Calif., private communication, August 24, 1993.

28. DSM savings are a Worldwatch estimate based on Eric Hirst, "Managing Demand for Electricity: Will It Pay Off?" *Forum for Applied Research and Public Policy*, Fall 1992, using an electricity price of 7.8¢ per kilowatt-hour (1993 cents) for 2010 from DOE, EIA, *Annual Energy Outlook 1994 With Projections to 2010* (Washington, D.C.: GPO, 1994), and assuming a DSM cost of 4¢ per kilowatt-hour.

29. EEI, "Pocketbook of Utility Statistics," op. cit. note 6.

30. Carl Weinberg, Joseph J. Iannucci, and Melissa M. Reading, "The Distributed Utility: Technology, Customer and Public Policy Changes Shaping the Electrical Utility of Tomorrow," PG&E Research and Development, San Ramon, Calif., December 1992; German law from Sara Knight, "Portrait of a Booming Market," *Windpower Monthly*, March 1993.

31. Henry Kelly and Carl J. Weinberg, "Utility Strategies for Using Renewables," in Thomas B. Johansson, Birgit Bodlund, and Robert H. Williams, eds., *Renewable Energy: Sources for Fuels and Electricity* (Washington, D.C.: Island Press, 1993).

32. Entergy from Steven R. Rivkin, "Look Who's Wiring the Home Now," *New York Times Magazine*, September 26, 1993.

33. Carl J. Weinberg and Katie McCormack, "Toward a Sustainable Energy Future," speech delivered by Weinberg to Consumer Federation of America, Washington, D.C., May 27, 1993; Leslie Lamarre, "The Vision of Distributed Generation," *EPRI Journal*, April/May 1993.

34. For a more detailed discussion of the status of the global elec-

tricity industry, see Christopher Flavin and Nicholas Lenssen, *Powering the Future: Blueprint for a Sustainable Electricity Industry,* Worldwatch Paper 119 (Washington, D.C.: Worldwatch Institute, June 1994); cost spread from Silverman, op. cit. note 14; "Global Electricity Prices," *Energy Economist,* September 1993; Tim Woolf, "Retail Competition in the Electricity Industry: Lessons from the United Kingdom," *The Electricity Journal,* June 1994.

35. Ralph C. Cavanagh, "The Great 'Retail Wheeling' Illusion—And More Productive Energy Futures," NRDC, San Francisco, Calif., draft, September 1993; Armond Cohen, senior attorney, Conservation Law Foundation, "Retail Wheeling and Rhode Island's Energy Future: Issues, Problems, and Lessons from Europe," presented to the Retail Wheeling Subcommittee of the Rhode Island Energy Coordinating Council, July 22, 1993.

36. Mark Ohrenschall, "Small Progressive Town Embraces Conservation," *Conservation Monitor,* November/December 1992; "German City Lashes Out at Global Climate Change," *Multinational Environmental Outlook,* September 18, 1990.

37. David H. Moskovitz, "Cutting the Nation's Electric Bill," *Issues in Science and Technology,* Spring 1989.

38. Among the many proposals for electric utility restructuring made recently, the ones closest to this are found in Hugh Outhred, University of New South Wales, "Achieving Least Cost Outcomes in the Emerging Competitive Electricity Industry," presented at the Second National Demand Management and Energy Efficiency Conference, Canberra, Australia, July 11-13, 1994, and in Stephen Wiel, "Achieving the Outcomes of Integrated Resource Planning within the Restructured Australian Electricity Industry," Lawrence Berkeley Laboratory, Washington, D.C., November 22, 1993.

39. Cohen, op. cit. note 35; Don Bain, "New Northwest Resources: Gas, or Conservation and Renewables?" Oregon Department of Energy, Salem, Oreg., December 23, 1992; Kelly and Weinberg, op. cit. note 31.

40. Pace University Center for Environmental Legal Studies, *Environmental Costs of Electricity* (New York: Oceana Publications, 1990); Jan Hamrin and Nancy Rader, *Investing in the Future: A Regulator's Guide to Renewables* (Washington, D.C.: National Association of Regulatory Utility Commissioners, 1993); David H. Moskovitz, "Green Pricing: Experience and Lessons Learned," The Regulatory Assistance Project, Gardiner, Maine, undated.

41. Silverman, op. cit. note 14.

42. Steven R. Rivkin and Jeremy D. Rosner, "Shortcut to the Information Superhighway: A Progressive Plan to Speed the Telecommunications Revolution," Policy Report No. 15, Progressive Policy Institute, Washington, D.C., July 1992.
43. S. David Freeman, general manager and chief executive officer, SMUD, Sacramento, Calif., private communication, October 13, 1993.
44. Matthew L. Wald, "Cuomo Appoints a New Head of New York Power Authority," *New York Times*, January 26, 1994.

CHAPTER 13. Through the Looking Glass

1. Peter Schwartz, *The Art of the Long View* (New York: Doubleday, 1991).
2. Ibid.
3. Ibid.
4. Daniel Yergin, Cambridge Energy Research Associates, Cambridge, Mass., private communication, 1984.
5. World Energy Council (WEC), *Energy for Tomorrow's World* (New York: St. Martin's Press, 1993).
6. Ibid.; Gerhard Ott, chairman, World Energy Council, "Energy for Tomorrow's World," speech to René Dubos Center for Human Environments Forum: World Energy to the Year 2020, New York, May 18–20, 1994.
7. Charles Darwin, *On the Origin of Species* (Cambridge, Mass.: Harvard University Press, 1964); N. Eldridge and S.J. Gould, "Punctuated Equilibria: An Alternative to Phyletic Gradualism," in T.J.M. Schopf, ed., *Models in Paleobiology* (San Francisco, Calif.: Freeman, Cooper and Co., 1972), cited in Stephen Jay Gould, *Hen's Teeth and Horse's Toes: Further Reflections in Natural History* (New York: W.W. Norton & Company, 1983).
8. "World Status: The Climate Change Treaty," *Energy Economist*, June 1992; Brian Barnett and W. Peter Teagan, Arthur D. Little, International, Cambridge, Mass., "R&D Funding for Fuel Cell Technology Development: Comparison with Other Advanced Power Technologies," prepared for World Fuel Cell Council, Frankfurt, December 1992.
9. Figure 13–1 is the result of a simple 5-parameter Worldwatch model tuned to reproduce carbon dioxide concentration projections done by the Intergovernmental Panel on Climate Change (IPCC) for its IS92a emissions scenario and by WEC for its B scenario. (The IS92a scenario is one of six done by IPCC; the other five project lower carbon emissions.) For all scenarios, carbon emissions from nonenergy sources (land use changes

and cement production) are taken from William Pepper et al., "Emission Scenarios for the IPCC: An Update," prepared for the IPCC Working Group I, May 1992. Historical emissions data are from R.M. Rotty and G. Marland, "Production of CO_2 From Fossil-Fuel Burning" (electronic database) (Oak Ridge, Tenn.: Oak Ridge National Laboratory (ORNL), 1993), and from G. Marland and T.A. Boden, "Global, Regional, and National CO_2 Emission Estimates from Fossil Fuel Burning, Cement Production, and Gas Flaring: 1950–1990" (electronic database) (Oak Ridge, Tenn.: ORNL, 1993). Scenarios are from WEC, op. cit. note 5, and from Pepper et al., op. cit. in this note. Given the tremendous scientific uncertainties involved in even sophisticated climate modelling, none of the results should be viewed as precise predictions.

10. Figures 13–2 and 13–3 are based on Pepper et al., op. cit. note 9, on WEC, op. cit. note 5, on Thomas B. Johansson et al., "A Renewables-intensive Global Energy Scenario" (Appendix to Chapter 1), in Thomas B. Johansson, Birgit Bodlund, and Robert H. Williams, eds., *Renewable Energy: Sources for Fuels and Electricity* (Washington, D.C.: Island Press, 1993) (for U.N. Solar Energy Future), and on Worldwatch scenario.

11. United Nations (UN), *Long-range World Population Projections: Two Centuries of Population Growth, 1950–2150* (New York: 1992); Paul Lewis, "U.N. Conference to Discuss Plan to Stabilize World Population," *New York Times*, April 3, 1994. Although population is an important variable in determining future carbon emission levels, rates of energy consumption both in industrial and developing countries will have a far greater effect.

12. Worldwatch estimate based on Sam H. Schurr and Bruce C. Netschert, *Energy in the American Economy, 1850–1975* (Baltimore, Md.: John Hopkins Press, 1960), on U.S. Department of Energy (DOE), Energy Information Agency (EIA), *Annual Energy Review* (Washington, D.C.: U.S. Government Printing Office (GPO), various years), on DOE, EIA, *Monthly Energy Review March 1994* (Washington, D.C.: GPO, 1994), on Nathan S. Balke and Robert J. Gordon, "The Estimation of Prewar Gross National Product: Methodology and New Evidence," Reprint No. 1200, National Bureau of Economic Research, Cambridge, Mass., and on Eunice Blue and Mary Perkins, U.S. Department of Commerce, Bureau of Economic Analysis, Washington, D.C., private communication and printout, April 8, 1993, and April 4, 1994.

13. Per capita energy use is based on British Petroleum (BP), *BP Statistical Review of World Energy* (London: 1993) and electronic

database (London: 1992), on D.O. Hall, King's College London, private communication, February 18, 1994, and printout, March 7, 1994, and on UN, *World Population Prospects: The 1992 Revision* (New York: 1993); Angus Maddison, "Growth and Slowdown in Advanced Capitalist Economies: Techniques of Quantitative Assessment," *Journal of Economic Literature*, June 1987.

14. WEC, op. cit. note 5; Nicholas Lenssen, "Nuclear Power Climbs," in Lester Brown, Hal Kane, and David Malin Roodman, *Vital Signs 1994* (New York: W.W. Norton & Company, 1994); Greenpeace International, WISE-Paris, and Worldwatch Institute, *The World Nuclear Industry Status Report: 1992* (London: Greenpeace International, 1992).

15. Mark Crawford, "Fusion Panel Drafts a Wish List for the '90s," *Science*, July 13, 1990; European Parliament (EP), Scientific and Technological Options Assessment (STOA), *Study on European Research into Controlled Thermonuclear Fusion* (Luxembourg: July 1991); see also Gordon J. Lake, STOA, EP, "Repatriating Refugees from Reality: Fusion and Fantasy in the European Community," presented at the Joint 4S/EASST 1992 Conference, Gothenburg, Sweden, August 12–15, 1992.

16. Biomass and hydropower shares are Worldwatch estimates, based on BP, op. cit. note 13, and on Hall, op. cit. note 13.

17. Capacity calculations assume a 50-percent conversion efficiency from primary energy in exajoules to kilowatt-hours of electricity generation, and a 30-percent capacity factor for both wind and solar thermal generators; also, we assume a 33-percent efficient conversion from primary energy to end-use electricity currently, rising to 50 percent in 2025 as thermal power plants become more efficient. In other words, a given amount of primary electricity supply in 2025 provides roughly half again as much end-use energy as it would currently.

18. Growth rate in nuclear capacity from R. Spiegelberg, Division of Nuclear Power, International Atomic Energy Agency, Vienna, private communication and printout, March 18, 1992; growth rate in personal computers from John E. Young, *Global Network: Computers in a Sustainable Society*, Worldwatch Paper 115 (Washington, D.C.: Worldwatch Institute, September 1993); size of Exxon from "The Fortune 500: The Largest U.S. Industrial Corporations," *Fortune*, April 18, 1994. The figure of $200 billion is based on assuming capacity factors of 30 percent for solar thermal and wind and 25 percent for photovoltaics, a 50-percent conversion ratio from primary to end-use energy, and annual capacity shipments (at $750 per kilowatt) sufficient to

raise installed capacity levels from 2025 to 2050 levels in the scenario; it includes operating costs of 0.5¢ per kilowatt-hour for the installed base in 2025.

19. Worldwatch estimate based on U.S. data from DOE, EIA, *Annual Energy Outlook 1994 With Projections to 2010* (Washington, D.C.: GPO, 1994), from DOE, EIA, *Monthly Energy Review September 1993* (Washington, D.C.: GPO, 1993), from Mohammad Adra, EIA, DOE, Washington, D.C., private communication, March 7, 1994, from DOE, EIA, *Household Energy Consumption and Expenditures 1990* (Washington, D.C.: GPO, 1993), and from Mueller Energy Technology Group, "Market Potential for Solar Thermal Energy Supply Systems in the United States Industrial and Commercial Sectors: 1990–2030," Arlington, Va., 1991; estimate counts only 20 percent of fuel used in road vehicles and 33 percent in other vehicles as "delivered."

20. Lawrence Fishbein and Carol J. Henry, "Health Effects of Methanol: An Overview," in Wilfrid L. Kohl, ed., *Methanol as an Alternative Fuel Choice: An Assessment* (Baltimore, Md.: Johns Hopkins University Press, 1990); Michael Lazarus et al., *Towards a Fossil Free Energy Future: The Next Energy Transition*, A Technical Analysis for Greenspeace International. (Boston: Stockholm Environment Institute, 1993); Thomas B. Johansson et al., "Renewable Fuels and Electricity for a Growing World Economy," in Johansson, Bodlund, and Williams, op. cit. note 10; Lester R. Brown and Hal Kane, *Full House: Reassessing the Earth's Population Carrying Capacity* (New York: W.W. Norton & Company, 1994).

21. Figure 13–4 is based on Schurr and Netschert, op. cit. note 12, on DOE, EIA, *Annual Energy Review 1992* (Washington, D.C.: GPO, 1992), and on DOE, *Monthly Energy Review March 1994*, op. cit. note 12.

22. Combusting hydrogen in the presence of air also produces nitrogen oxides, though the amount produced varies widely depending on the combustion process.

23. Verne reference from Peter Hoffmann, *The Forever Fuel: The Story of Hydrogen* (Boulder, Colo.: Westview Press, 1981).

24. Joan M. Ogden and Joachim Nitsch, "Solar Hydrogen," in Johansson, Bodlund, and Williams, op. cit. note 10; 1 percent figure is a Worldwatch estimate, based on DOE, *Monthly Energy Review March 1994*, op. cit. note 12, on DOE, EIA, *State Energy Data Report 1991: Consumption Estimates* (Washington, D.C.: GPO, 1993), and on Wayne B. Solley et al., *Estimated Use of Water in the United States in 1990*, United States Geological Survey Circular 1081 (Washington, D.C.: GPO, 1993).

25. Derek P. Gregory, "The Hydrogen Economy," *Scientific American*, January 1973.

26. Electrolysis units have only limited economies of scale above 2 megawatts, according to Joan M. Ogden and Robert H. Williams, *Solar Hydrogen: Moving Beyond Fossil Fuels* (Washington, D.C.: World Resources Institute, 1989).

27. Ogden and Nitsch, op. cit. note 24.

28. Philip H. Abelson, "Applications of Fuel Cells" (editorial), *Science*, June 22, 1990.

29. Ogden and Nitsch, op. cit. note 24; loss of 3–5 percent is a Worldwatch estimate based on Hydro-Quebec's operating experience with both alternating and direct current long-distance power transmission lines, from Raymond Elsliger, Hydro-Quebec, Montreal, private communication, May 24, 1994.

30. BP, op. cit. note 13.

31. Table 13–1 is based on Keith Lee Kozloff and Roger C. Dower, *A New Power Base: Renewable Energy Policies for the Nineties and Beyond* (Washington, D.C.: World Resources Institute, 1993), on Paul Gipe, Gipe and Associates, Tehachapi, Calif., private communication and printout, March 29, 1994, and on David Mills, University of Sydney, Australia, private communication and printout, March 25, 1994.

32. Land area calculations based on data in Table 13–1, assuming a 50-percent conversion efficiency from primary energy in exajoules to kilowatt-hours of electricity generation, and a 30-percent capacity factor for both wind and solar thermal generators; the land needed for solar and wind power in 2100 under our scenario is less than double the 2050 figures.

33. Population from UN, op. cit. note 11; carbon figures are Worldwatch estimates, based on Joel Darmstadter, Perry D. Teitelbaum and Jaroslav G. Polach, *Energy in the World Economy* (Baltimore, Md.: Johns Hopkins University Press, 1971), on Rotty and Marland, op. cit. note 9, on Marland and Boden, op. cit. note 9, and on BP, op. cit. note 13.

34. Worldwatch projections.

CHAPTER 14. Launching the Revolution

1. United Nations, Report of the Intergovernmental Negotiating Committee for a Framework Convention on Climate Change, Fifth session, Second part, New York, April 30-May 9, 1992.

2. Interim Secretariat of the United Nations Framework Convention on Climate Change, "Status of Ratification of the United Nations Framework Convention on Climate Change" (electronic bulletin board posting), April 15, 1994.

3. Ibid.
4. Carl J. Weinberg and Robert H. Williams, "Energy from the Sun," *Scientific American*, September 1990.
5. Fossil fuel subsidies do not subtract revenue from existing taxes on fuels, and are from Bjorn Larsen, "World Fossil Fuel Subsidies and Global Carbon Emissions in a Model with Interfuel Substitution," World Bank, Washington, D.C., February 1994; tariffs in developing countries from Energy Development Division, *Review of Electricity Tariffs in Developing Countries During the 1980's*, Industry and Energy Department Working Paper, Energy Series Paper No. 32 (Washington, D.C.: World Bank, 1990).
6. Douglas N. Koplow, *Federal Energy Subsidies: Energy, Environmental, and Fiscal Impacts* (Washington, D.C.: Alliance to Save Energy, 1993); "Nuclear Electric Eyes No. 2 Spot," *European Energy Report*, August 6, 1993; Organisation for Economic Co-operation and Development (OECD), International Energy Agency (IEA), *Energy Policies of IEA Countries: 1992 Review* (Paris: 1993); Mark Clayton, "Analysis Decries 1980s Dealmaking by Quebec's Hydro-Electric Utility," *Christian Science Monitor*, March 14, 1994.
7. Worldwatch estimate based on British Petroleum (BP), *BP Statistical Review of World Energy* (London: 1993), and on Robert Chiarotti, World Coal Institute, London, private communication and printout, February 22, 1994.
8. Ernst U. von Weizsäcker and Jochen Jesinghaus, *Ecological Tax Reform* (Atlantic Highlands, N.J.: Zed Books, 1992).
9. Per Kågeson, *Getting Prices Right: A European Scheme for Making Transport Pay its True Costs* (Stockholm: European Federation for Transport and Environment, 1993); receipts from carbon tax is from Worldwatch estimate of emissions based on Thomas A. Boden, Oak Ridge National Laboratory (ORNL), Oak Ridge, Tenn., private communication and printout, September 20, 1993, and on BP, op. cit. note 7; G. Marland and T.A. Boden, "Global, Regional, and National CO_2 Emission Estimates from Fossil Fuel Burning, Cement Production, and Gas Flaring: 1950–1990" (electronic database) (Oak Ridge, Tenn.: ORNL, 1993); Bjorn Larsen and Anwar Shah, "World Fossil Fuels Subsidies and Global Carbon Emissions," World Bank, February 20, 1992; Roger C. Dower and M.B. Zimmerman, *The Right Climate for Carbon Taxes: Creating Economic Incentives to Protect the Atmosphere* (Washington, D.C.: World Resources Institute, 1992).
10. "EC Commission to Consider Proposals for Energy Taxes,"

Journal of Commerce, June 24, 1991; David Gardner, "Japan Shows 'Clear Interest' in Community's Energy Tax Plans," *Financial Times*, May 20, 1992; Swedish Environmental Protection Agency, "The Carbon Dioxide Tax and Energy Taxes— The Swedish Experience," Solna, undated; Norwegian Ministry of Environment, "Carbon Taxes: Norwegian Experiences," Oslo, August 1993; "Berne to Introduce CO_2 Tax Based on Fossil Fuel Emissions," *European Energy Report*, April 1, 1994; Delors cited in Lionel Barber, "Environment is Top of Clinton's Trade Agenda," *Financial Times*, January 12, 1994.

11. Robert W. Hahn and Carol A. May, "The Behavior of the Allowance Market: Theory and Evidence," *Electricity Journal*, March 1994; Matthew Wald, "2 Utilities Intend Interstate Trade of Smog Pollution," *New York Times*, March 16, 1994; Robert Reinhold, "Hard Times Dilute Enthusiasm for Clean-Air Laws," *New York Times*, November 26, 1993.

12. Table 14–1 is based on Brian Barnett and W. Peter Teagan, "R&D Funding for Fuel Cell Technology Development: Comparison with Other Advanced Power Technologies," Arthur D. Little International, Cambridge, Mass., prepared for World Fuel Cell Council, December 1992, on OECD, IEA, *Energy Policies of IEA Countries: 1991 Review* (Paris: 1992), and on Worldwatch estimates; Mark Crawford, "Fusion Panel Drafts a Wish List for the '90s," *Science*, July 13, 1990.

13. Fred J. Sissine, "Renewable Energy: A New National Commitment?" Congressional Research Service, Washington, D.C., January 5, 1994.

14. Boston Consulting Group research is cited in Robert H. Williams and Gregory Terzian, "A Benefit/Cost Analysis of Accelerated Development of Photovoltaic Technology," Center for Energy and Environmental Studies, Princeton University, Princeton, N.J., October 1993; Figure 14–1 is redrawn from William J. Abernathy and Kenneth Wayne, "Limits of the Learning Curve," *Harvard Business Review*, September-October 1974.

15. Wind turbine output is a Worldwatch estimate based on Paul Gipe, Paul Gipe and Associates, Tehachapi, Calif., private communication and printout, April 6, 1994; John Trocciola, International Fuel Cells, South Windsor, Conn., private communication, April 15, 1994; Figure 14–2 is a based on photovoltaic prices from Paul Maycock, Photovoltaic Energy Systems, Inc., Casanova, Va., private communications and printouts, May 8, 1992, and December 20, 1993, and on production data from Paul D. Maycock, "1993 World Module Shipments," *Photovol-*

taic News, February 1994, and from Paul Maycock, private communication, March 23, 1994.

16. "MITI to Subsidize Home Photovoltaics," *The Quad Report*, Washington, D.C., November 1993; Steven J. Strong, "An Overview of Worldwide Development Activity in Building-integrated Photovoltaics," Solar Design Associates, Inc., Harvard, Mass., undated; Utility PhotoVoltaic Group, "Electric Utilities Serving 40% of U.S. Consumers Propose $513 Million Program to Accelerate Use of Solar Photovoltaics," Washington, D.C., September 27, 1993; Ann Polansky, Solar Energy Industries Association, private communication, April 26, 1994.

17. John DeCicco and Deborah Gordon, American Council for an Energy-Efficient Economy (ACEEE), "Steering with Prices: Fuel and Vehicle Taxation as Market Incentives for Higher Fuel Economy," presented at the Transportation and Energy Conference, Asilomar, Calif., August 22–25, 1993, and updated, December 19, 1993; John DeCicco et al., "Feebates for Fuel Economy: Market Incentives for Encouraging Production and Sales of Efficient Vehicles," ACEEE, Washington, D.C., 1993; Robin Miles-McLean, Susan M. Haltmaier, and Michael G. Shelby, "Designing Incentive-Based Approaches to Limit Carbon Dioxide Emissions from the Light-Duty Vehicle Fleet," in David L. Greene and Danilo J. Santini, *Transportation and Global Climate Change* (Washington, D.C.: ACEEE, 1993); European Parliament, "Report of the Committee on Transport and Tourism on Transport and Energy," Brussels, January 26, 1994.

18. Amulya K.N. Reddy and José Goldemberg, "Energy for the Developing World," *Scientific American*, September 1990.

19. John Besant-Jones, World Bank, Washington, D.C., private communication, August 24, 1992; Mark Levine et al., *Energy Efficiency, Developing Nations, and Eastern Europe*, A Report to the U.S. Working Group on Global Energy Efficiency (Washington, D.C.: International Institute for Energy Conservation (IIEC), 1991); World Bank, *Annual Report 1993* (Washington, D.C.: 1993); African Development Bank, *1990 Annual Report* (Abidjan: 1991); Asian Development Bank, *Annual Report 1991* (Manila: 1992); Inter-American Development Bank, *1991 Annual Report* (Washington, D.C.: 1992); Michael Philips, *The Least Cost Energy Path for Developing Countries: Energy Efficient Investments for the Multilateral Development Banks* (Washington, D.C.: IIEC, 1991).

20. World Bank, *The World Bank's Role in the Electric Power Sector* (Washington, D.C.: 1993); Environmental Defense Fund and

Natural Resources Defense Council, *Power Failure: A Review of the World Bank's Implementation of its New Energy Policy* (New York: Environmental Defense Fund, 1994).

21. Global Environment Facility (GEF), "Report by the Chairman to the April 1992 Participants' Meeting," Washington, D.C., March 1992; GEF, "A Selection of Projects from the First Three Tranches," Working Paper Series Number II, Washington, D.C., June 1992; GEF, "Mauritius Sugar Bio-Energy Technology Project," Washington, D.C., January 1992.

22. Maria Subiza, GEF, Washington, D.C., private communication, April 12, 1994; GEF, "Instrument for the Establishment of the Restructured Global Environment Facility," Washington, D.C., March 31, 1994; Philips, op. cit. note 19.

23. Megan Ryan, "Power Move: The Nuclear Salesmen Target Developing Countries," *World Watch*, March/April 1994.

24. United Nations, Committee on the Development and Utilization of New and Renewable Sources of Energy, "Commemoration of the Tenth Anniversary of the Adoption of the Nairobi Programme of Action for the Development and Utilization of New and Renewable Sources of Energy," Report of the Intergovernmental Group of Experts on New and Renewable Sources of Energy, February 3–14, 1992.

Index

ABOUT THE AUTHORS

CHRISTOPHER FLAVIN is Vice-President for Research at the Worldwatch Institute in Washington, D.C., a private, nonprofit environmental research organization. His research focuses on energy resource, technology, and policy issues. He is active in U.S. energy debates and in international discussions about climate change, nuclear power, and the development of new energy sources. He has also written on national and international economic issues, including the need for private corporations to adopt new approaches to environmental issues. Mr. Flavin is the author of 12 Worldwatch Papers and a contributor to Worldwatch's annual *State of the World* and *Vital Signs*. In 1992, he helped found the Washington-based Business Council for a Sustainable Energy Future, and serves on the board. Mr. Flavin is a graduate of Williams College, with a B.A. in economics, biology, and environmental studies.

NICHOLAS LENSSEN is a Senior Researcher at the Worldwatch Institute. His research and writing deals with energy policy, alternative energy sources, nuclear power, radioactive waste, and global climate change. He has testified before the U.S. Congress and the European Parliament on energy issues, written four Worldwatch Papers, and contributed to the Institute's two annual books. Mr. Lenssen is a graduate of Dartmouth College, with an A.B. in geography. Before joining Worldwatch, he worked at the National Wildlife Federation, the New Mexico State Senate, and the Rocky Mountain Institute and with the U.S. Peace Corps in Ecuador.